P9-APU-990

Why Aren't More Women in Science?

Why Aren't More Women in Science?

Top Researchers Debate the Evidence

Edited by
**Stephen J. Ceci and
Wendy M. Williams**

American Psychological Association • Washington, DC

Published by
American Psychological Association
750 First Street, NE
Washington, DC 20002
www.apa.org

To order
APA Order Department
P.O. Box 92984
Washington, DC 20090-2984
Tel: (800) 374-2721; Direct: (202) 336-5510
Fax: (202) 336-5502; TDD/TTY: (202) 336-6123
Online: www.apa.org/books/
E-mail: order@apa.org

In the U.K., Europe, Africa, and the Middle East, copies may be ordered from
American Psychological Association
3 Henrietta Street
Covent Garden, London
WC2E 8LU England

Typeset in Goudy by Stephen McDougal, Mechanicsville, MD

Printer: Edwards Brothers, Inc., Ann Arbor, MI
Cover Designer: Naylor Design, Washington, DC
Technical/Production Editor: Tiffany L. Klaff

The opinions and statements published are the responsibility of the authors, and such opinions and statements do not necessarily represent the policies of the American Psychological Association.

Library of Congress Cataloging-in-Publication Data

Ceci, Stephen J.
   Why aren't more women in science? : top researchers debate the evidence /
Stephen J. Ceci and Wendy M. Williams. — 1st ed.
       p. cm.
   Includes bibliographical references and index.
   ISBN-13: 978-1-59147-485-2
   ISBN-10: 1-59147-485-X
   1. Women in science. 2. Sex discrimination in science. I. Williams, Wendy M. (Wendy
Melissa), 1960– II. Title.

   Q130.C43 2007
   508.2—dc22   2006025963

British Library Cataloguing-in-Publication Data
A CIP record is available from the British Library.

Printed in the United States of America
First Edition

To our daughters: Sterling Chance, Nicole Genevieve, and Wynne Linnea

CONTENTS

CONTRIBUTORS

Simon Baron-Cohen, PhD, MPhil, is professor of developmental psychopathology at the University of Cambridge and fellow at Trinity College, Cambridge, England. He is director of the Autism Research Centre in Cambridge. He holds degrees in human sciences from New College, Oxford, England; a doctorate in psychology from University College London; and master of philosophy in clinical psychology at the Institute of Psychiatry. He is the author of *Mindblindness* (1995), *The Essential Difference: Men, Women and the Extreme Male Brain* (2003), and *Prenatal Testosterone in Mind* (2005). He has been awarded prizes from the American Psychological Association and the British Psychological Society for his research on autism.

Camilla Persson Benbow, EdD, is a professor of psychology and human development and the Patricia and Rodes Hart Dean of the Peabody College of Education and Human Development at Vanderbilt University in Nashville, Tennessee. She completed all her degrees at Johns Hopkins University in Baltimore, Maryland. Her scholarship has concentrated predominantly on intellectually talented children, with her research program seeking to identify different "types" of intellectually talented adolescents, characterize them, and then discover effective ways to facilitate their development. Through David Lubinski's and her longitudinal study of over 5,000 individuals (the Study of Mathematically Precocious Youth), she examines their developmental trajectories and the impact of educational interventions over the life span. For her work, she has received numerous awards, most recently the Mensa Education and Research Foundation's Lifetime Achievement Award (2004).

Sheri A. Berenbaum, PhD, is professor of psychology and pediatrics and a member of the Neuroscience Institute at The Pennsylvania State University. Her research focuses on genetic and neuroendocrine influences on the development of human cognition and social behavior. She is particularly interested in the effects of prenatal sex hormones on the development of sex-typed behaviors and how these effects are mediated directly by the brain and indirectly through the social environment. Her behavioral studies of children and adults exposed to high prenatal levels of androgens have been supported by the National Institute of Child Health and Human Development since 1985. She also leads a research network on psychosexual differentiation funded by the National Institutes of Health (NIH). She has been on the faculty at the University of Health Sciences/Chicago Medical School and the Southern Illinois University School of Medicine, where she received the 1999 Faculty University Woman of Distinction Award. She is a fellow of the American Psychological Association, served on an NIH study section, and is a present or past member of several editorial boards and professional committees, including those concerned with treatment of children with ambiguous genitalia. She was a member of the Institute of Medicine committee that wrote the report "Exploring the Biological Contributions to Human Health: Does Sex Matter?"

Stephen J. Ceci, PhD, Helen L. Carr Professor of Developmental Psychology at Cornell University, has published extensively on topics related to group differences in achievement, intellectual development, and gap-closing. He is a member of the White House Commission on Children; the National Academy of Sciences Board on Cognitive, Sensory, and Behavioral Sciences since 1996; the National Science Foundation Advisory Board since 1997; and various National Research Council committees. He is a fellow of the American Psychological Association, the Association for Psychological Science, and the American Association for the Advancement of Science. Dr. Ceci founded and coedits the journal *Psychological Science in the Public Interest*, published by the Association for Psychological Science. In 2003 Dr. Ceci was the corecipient of the American Psychological Association's Lifetime Award for the Scientific Application of Psychology (with Elizabeth F. Loftus), and in 2005 he was the corecipient of the Association for Psychological Science's James McKeen Cattell Fellow Award for lifetime contribution (with E. Mavis Hetherington). He has published over 300 articles, chapters, books, and reviews, including several citation classics and 14 articles and books cited over 100 times each. In both 1997 and 2002 he was listed by Byrne and McNamara (*Developmental Review*) as one of the most cited developmental psychologists.

Carol S. Dweck, PhD, received her doctorate from Yale University in 1972 and is the Lewis and Virginia Eaton Professor of Psychology at Stanford University. For over 30 years, her research has been devoted to identifying the foundations of motivation and to explaining why some people fulfill their potential whereas others of equal ability do not. Through this motivational approach, she has contributed to our understanding of such things as gender differences in achievement, the impact of stereotypes on performance, and the neural correlates of effective and ineffective learning. She has also investigated the developmental roots of motivational patterns and the socialization practices that foster them. Finally, on the basis of her work, she has developed an intervention that consistently changes students' motivation and boosts their achievement. She has received numerous honors and awards and was elected to the American Academy of Arts and Sciences. Her books include *Motivation and Self-Regulation Across the Lifespan* (with Jutta Heckhausen, 1998); *Self-Theories: Their Role in Motivation, Personality, and Development* (2000); *The Handbook of Competence and Motivation* (with Andrew Elliot, 2005); and *Mindset* (2006).

Jacquelynne S. Eccles, PhD, is McKeachie Collegiate Professor of Psychology at the University of Michigan, Ann Arbor. She received her doctorate from the University of California, Los Angeles, in 1974 and has served on the faculty at Smith College, the University of Colorado, and the University of Michigan. She has chaired the MacArthur Foundation Network on Successful Pathways Through Middle Childhood and was a member of the MacArthur Research Network on Successful Pathways Through Adolescence. She was program chair for the Society for Research on Adolescence (SRA), has served on the SRA Council, and is now past president of SRA. She is currently a member of the National Institute of Child Health and Human Development governing council and chaired the National Academy of Sciences Committee on Community-Based Programs for Youth. Her awards include the Spencer Foundation Fellowship for Outstanding Young Scholar in Educational Research, the Sarah Goddard Power Award for Outstanding Service from the University of Michigan, the American Psychological Society Cattell Fellows Award for Outstanding Applied Work in Psychology, the Society for the Psychological Study of Social Issues Kurt Lewin Award for outstanding research, and the Thorndike Career Achievement Award from Division 15 of the American Psychological Association. She has conducted research on topics ranging from gender role socialization, self-related beliefs and achievement-related choices, and classroom influences on motivation to social development in the family, school, peer, and wider cultural contexts.

David C. Geary, PhD, received his doctorate in developmental psychology in 1986 from the University of California, Riverside, and held faculty positions at the University of Texas at El Paso and the University of Missouri, first at the Rolla campus and then in Columbia. He served as chair of the Department of Psychological Sciences from 2002 to 2005, as the University of Missouri's Middlebush Professor of Psychological Sciences from 2000 to

2003, and is now a Curators' Professor. He has published nearly 150 articles, commentaries, and chapters across a wide range of topics, including cognitive and developmental psychology, education, evolutionary biology, and medicine. His three books are *Children's Mathematical Development* (1994); *Male, Female: The Evolution of Human Sex Differences* (1998); and *The Origin of Mind: Evolution of Brain, Cognition, and General Intelligence* (2005), all published by the American Psychological Association. He has given invited addresses in a variety of departments (anthropology, biology, behavior genetics, computer science, education, government, mathematics, neuroscience, physics, and psychology) and universities throughout the United States as well as in Austria, Belgium, Canada, Germany, and Italy. He was one of the primary contributors to the Mathematics Framework for California Public Schools: Kindergarten Through Grade Twelve and is currently a member of the President's National Mathematics Panel. Among his many distinctions are the Chancellor's Award for Outstanding Research and Creative Activity in the Social and Behavioral Sciences (1996) and a MERIT Award from the National Institutes of Health.

Ariel D. Grace, BA, graduated from Oberlin College with a bachelor's degree in psychology in 2001. She is currently a PhD student in cognitive psychology at Northwestern University in Evanston, Illinois.

Raquel E. Gur, MD, PhD, is Karl and Linda Rickels Professor of Psychiatry, with appointments in neurology and radiology at the University of Pennsylvania in Philadelphia. Her academic career has been devoted to the study of brain function in schizophrenia. After completing her PhD at Michigan State University, in East Lansing, in 1974, she joined the University of Pennsylvania, first as a fellow (1974), then as faculty (1976), medical student (1977), and resident in neurology (1980) and psychiatry (1984). She has directed the Neuropsychiatry Section and the Schizophrenia Research Center and has established an interdisciplinary program dedicated to advancing the understanding of the pathophysiology of this complex disorder through the application of diverse strategies, from neurobehavioral to molecular. The research has been supported by the National Institute of Mental Health, where she is currently serving on the National Advisory Mental Health Council. She has over 200 refereed publications, is action editor of *Neuropsychopharmacology*, and her accomplishments have been acknowledged by awards and membership in the Institute of Medicine. She is president elect of the Society of Biological Psychiatry.

Ruben C. Gur, PhD, received his bachelor's degree in psychology and philosophy from the Hebrew University of Jerusalem, Israel, in 1970 and his doctorate in psychology (clinical) from Michigan State University, in East Lansing, in 1973. After postdoctoral training with E. R. Hilgard at Stanford

University, in Stanford, California, he taught at the University of Pennsylvania in Philadelphia as assistant professor in 1974, where he is currently professor of psychology in the psychiatry, neurology, and radiology departments and director of the Brain Behavior Laboratory and the Center for Neuroimaging in Psychiatry. His research has been in the study of brain and behavior in healthy people and patients with brain disorders, with a special emphasis on exploiting neuroimaging as experimental probes. His work has documented sex differences, aging effects, and abnormalities in regional brain function associated with schizophrenia, affective disorders, stroke, epilepsy, movement disorders, and dementia. He is recipient of the 1990 Stephen V. Logan Award, National Alliance for the Mentally Ill. He has over 200 refereed publications and is action editor of *Brain and Cognition*. He serves on a National Institute of Mental Health Study Section on Neural Basis of Psychopathology, Addictions and Sleep Disorders, and his work has been supported by grants from the National Science Foundation, National Institutes of Health, National Institute of Mental Health, National Institute on Aging, National Institute of Neurological Disorders and Stroke, Department of Defense, private foundations (Spencer, MacArthur, the EJLB Foundation), and industry (Merck, Lilly, Pfizer).

Richard J. Haier, PhD, is professor of psychology in the Department of Pediatrics and the Department of Psychology and Social Behavior at the University of California, Irvine. He received his doctorate from Johns Hopkins University and has had appointments at the National Institute of Mental Health (Laboratory of Psychology and Psychopathology) and at Brown University (Department of Psychiatry and Human Behavior). As one of the first psychologists to use positron emission tomography (PET) in the 1980s, he has published numerous research papers addressing individual differences in intelligence, learning, and personality. These include some of the first brain imaging studies of sex differences as they relate to cognition. Recently, he reported that men and women may differ with respect to the structural brain areas associated with general intelligence. His most recent funding from the National Institutes of Health addresses the sequence and progression of structural and functional brain changes in preclinical dementia.

Diane F. Halpern, PhD, is professor of psychology and director of the Berger Institute for Work, Family, and Children at Claremont McKenna College. She was president (2004) of the American Psychological Association (APA). Dr. Halpern has won many awards for her teaching and research, including the 2002 Outstanding Professor Award from the Western Psychological Association, the 1999 American Psychological Foundation Award for Distinguished Teaching, the 1996 APA Distinguished Career Award for Contributions to Education, the California State University's State-Wide Outstanding Professor Award, the Outstanding Alumna Award from the University of

Cincinnati, the Silver Medal Award from the Council for the Advancement and Support of Education, the Wang Family Excellence Award, and the G. Stanley Hall Lecture Award from the APA. She is the author of many books: *Thought and Knowledge: An Introduction to Critical Thinking; Thinking Critically About Critical Thinking* (with Heidi Riggio); *Sex Differences in Cognitive Abilities; Enhancing Thinking Skills in the Sciences and Mathematics; Changing College Classrooms; Student Outcomes Assessment;* and *States of Mind: American and Post-Soviet Perspectives on Contemporary Issues in Psychology* (coedited with Alexander Voiskounsky). Her most recent book is coedited with Susan Murphy, titled *From Work–Family Balance to Work–Family Interaction: Changing the Metaphor.*

Melissa Hines, PhD, is professor of psychology and director of the Behavioural Neuroendocrinology Research Centre at City University, London, England. She earned her doctorate in psychology in 1981 from the University of California, Los Angeles (UCLA), and did postdoctoral training in neuroendocrinology at the UCLA Brain Research Institute. She is also a licensed clinical psychologist in California and a chartered counseling psychologist in Great Britain. She is the past president of the International Academy of Sex Research and serves on the editorial boards of the journals *Hormones and Behavior* and *Archives of Sexual Behavior.* She is the author of the book *Brain Gender* (2004). Currently, she directs two major research projects. The first, funded by the Wellcome Trust, is a longitudinal study examining how prenatal hormones and postnatal socialization combine to influence gender development in a normal population sample. The second, funded by the U.S. Public Health Service, National Institutes of Health, is examining the social, cognitive, and neural mechanisms by which prenatal hormones influence postnatal behavior. This research program includes studies of individuals born with disorders of sex development (also called intersex conditions), as well as studies of healthy children for whom prenatal hormone measurements are available. Both research projects are looking at a range of sex-typed behaviors, including gender role behaviors, cognitive outcomes, and gender-related occupational and other interests.

Janet Shibley Hyde, PhD, is Helen Thompson Woolley professor of psychology and women's studies at the University of Wisconsin—Madison. She earned her doctorate in 1972 from the University of California, Berkeley. She is the author of a textbook for the psychology of women course titled *Half the Human Experience: The Psychology of Women.* One line of her research has focused on gender differences in abilities and self-esteem. Another line focuses on women, work, and dual-earner couples. One current research project, the Wisconsin Maternity Leave and Health Project (now called the Wisconsin Study of Families and Work), focuses on working mothers and their children. Another current project, funded by the National Science

Foundation, is the Moms & Math Project, in which she is studying mothers interacting with their fifth- or seventh-grade children as they do mathematics homework together. Other research investigates the emergence of gender differences in depression in adolescence and peer sexual harassment in adolescence. She is a fellow of the American Psychological Association and the American Association for the Advancement of Science, and a winner of the Heritage Award from the Society for the Psychology of Women for career contributions to research on the psychology of women and gender.

Doreen Kimura, PhD, FRSC, LLD (Hon), received both undergraduate and graduate degrees at McGill University, Montreal, Quebec, Canada, and spent postdoctoral years at the University of California, Los Angeles Medical Center and at the Neurosurgical Clinic in the Kantonspital, Zurich, Switzerland. She was on the faculty at the University of Western Ontario, Ontario, Canada, for over 30 years and currently has a postretirement visiting professorship in psychology at Simon Fraser University, British Columbia, Canada. Her research interest has been in the biological mechanisms in human cognitive function, including the influence of individual differences such as sex, sexual orientation, and handedness. She has studied the role of the left- and right-brain hemispheres in language, complex motor function, semantic function, auditory processing, and nonverbal memory. She has also studied human sex differences in cognitive function and the hormonal influences that mediate some of those differences, and she has researched the association of cognitive patterns with various body asymmetries. She is the author of two academic books: *Neuromotor Mechanisms in Human Communication* (1993) and *Sex and Cognition* (1999, paperback 2000). The latter has been translated into several languages. She is a past fellow of the Canadian Psychological Association and American Psychological Association, has received honorary degrees from Simon Fraser University and Queen's University, Kingston, Ontario, Canada, and is a fellow of the Royal Society of Canada. Her most recent distinction was the Hebb Award in 2005 from the Canadian Society for Brain, Behaviour and Cognitive Science. In addition to her research interests, she was the founding president of the Society for Academic Freedom and Scholarship, received the Furedy Award in 2002 from that society for her support of academic freedom, and has published a collection of her writings on those issues in *Dissenting Opinions* (2002).

David S. Lubinski, PhD, received both his BA (1981) and PhD (1987) in psychology from the University of Minnesota, Twin Cities. From 1987 to 1990 he was a fellow in the Postdoctoral Training Program in Quantitative Methods, Department of Psychology, University of Illinois at Urbana–Champaign. He is currently professor of psychology and human development at Vanderbilt University. With Camilla Persson Benbow, he codirects the Study of Mathematically Precocious Youth, a planned 50-year longitudinal study of over 5,000 intellectually talented participants, begun in 1971 (under Julian C. Stanley). His work has earned him the American Psychological Association (APA) 1996 Distinguished Scientific Award for Early Career Contribution to Psychology (Applied Research/Psychometrics), APA's 1996 George A. Miller Award (Outstanding Article in General Psychology), the 1995 American Educational Research Association's Research Excellence Award (Counseling/Human Development), six Mensa awards for research excellence, APA's Templeton Award (2000) for Positive Psychology, the Cattell Sabbatical Award (2003–2004), and the National Association for Gifted Children's Distinguished Scholar Award (2006). He is a fellow of APA and the American Psychological Society, is a member of the Society for Multivariate Experimental Psychology, and serves on the advisory board for the International Society for Intelligence Research. He has served as associate editor, *Journal of Personality and Social Psychology: Personality Processes and Individual Differences*, and in a recent *Developmental Review* study, his productivity was rated in the top 5% among developmental science faculty.

Nora S. Newcombe, PhD, is professor of psychology and James H. Glackin Distinguished Faculty Fellow at Temple University. She has also been a visiting scholar at the University of Pennsylvania, at Princeton University (supported by a Cattell Fellowship), and at the Wissenschaftskolleg in Berlin, Germany. She received her doctorate in psychology and social relations from Harvard University. Her research focuses on spatial development and the development of episodic and autobiographical memory. She has served as editor of the *Journal of Experimental Psychology: General* and as associate editor of *Psychological Bulletin*, as well as on the Human Cognition and Perception Panel at the National Science Foundation and many editorial boards. She is the author of numerous scholarly chapters and articles on aspects of cognitive development and the author or editor of three books, including *Making Space: The Development of Spatial Representation and Reasoning* (with Janellen Huttenlocher) in 2000. Her work has been recognized by the George A. Miller Award from the American Psychological Association (APA) and by the Paul W. Eberman Research Award from Temple University. She is a fellow of four divisions of the American Psychological Association (Society for General Psychology, Experimental Psychology, Developmental Psychology, and Society for the Psychology of Women), of the American Psychological Society, and of the American Association for the Advancement of Science.

Susan Resnick, PhD, is senior investigator in the Laboratory of Personality and Cognition at the National Institute on Aging (NIA). She received her doctorate in differential psychology and behavioral genetics from the University of Minnesota in 1983. She completed a postdoctoral fellowship in neuropsychology and neuroimaging at the University of Pennsylvania, where she was research assistant professor prior to joining the NIA in 1992. She is a member of the editorial board of *Brain and Cognition* and a frequent reviewer for many medical, psychological, and neuroimaging journals and the National Institutes of Health Office of Research on Women's Health. She studies brain–behavior associations in health and disease and factors that modify cognitive and brain aging, including sex differences and sex steroid hormones. She is the principal investigator of the brain imaging substudy of the Baltimore Longitudinal Study of Aging, a longitudinal neuroimaging study focusing on brain changes as early markers of cognitive impairment in the elderly. Through this study and others, she investigates hormonal modulation of age-associated cognitive and brain changes. She played a key role in the development of the Women's Health Initiative Study of Cognitive Aging (WHISCA) and Cognition in the Study of Tamoxifen and Raloxifene (Co-STAR), which investigate effects of menopausal hormone therapy and selective estrogen receptor modulators, respectively, on cognitive aging in postmenopausal women. She is also a member of the planning group for a multisite clinical trial of testosterone supplementation in older hypogonadal men.

Elizabeth S. Spelke, PhD, teaches at Harvard University, Cambridge, Massachusetts, where she is the Marshall L. Berkman Professor of Psychology and codirector of the Mind, Brain, and Behavior Initiative. She studies the origins and nature of knowledge of objects, persons, space, and numbers through research on human infants, children, human adults in diverse cultures, and nonhuman animals. A member of the National Academy of Sciences and the American Academy of Arts and Sciences, and cited by *Time* magazine as one of America's Best in Science and Medicine, her honors include the Distinguished Scientific Contribution Award of the American Psychological Association, the William James Award of the American Psychological Society, and honorary degrees from the University of Umea, Sweden, and the École Pratique des Hautes Études, Paris, France.

Virginia Valian, PhD, is distinguished professor of psychology and linguistics at Hunter College and the Graduate Center of the City University of New York. She is a cognitive scientist whose research ranges from first- and second-language acquisition to gender equity. The National Science Foundation (NSF) currently funds both her research in language acquisition and her work on gender. Dr. Valian is codirector of Hunter College's Gender Equity Project (partially funded by an NSF ADVANCE Institutional Transformation Award), the initiatives of which are described at http://www.hunter.cuny.edu/genderequity. She has also created Web-based tutorials (funded by the NSF) on the role of gender in professional life. These tutorials (http://www.hunter.cuny.edu/gendertutorial) take the form of slides with voiceover narration and are intended for students, faculty, and administrators worldwide. Her book, *Why So Slow? The Advancement of Women*, published in 1998, has been described by reviewers as "compelling," "scholarly and convincing," "accessible and lively," and "a breakthrough in the discourse on gender." Dr. Valian lectures and gives workshops to faculty and professional groups in the United States and abroad.

Wendy M. Williams, PhD, is professor in the Department of Human Development at Cornell University, where she studies the development, assessment, training, and societal implications of intelligence and related abilities. She holds a doctorate and master's degree in psychology from Yale University; a master's degree in physical anthropology from Yale; and a bachelor's degree in English and biology from Columbia University, which she was awarded cum laude with special distinction. Dr. Williams cofounded and codirects the Cornell Institute for Research on Children, a National Science Foundation–funded research- and outreach-based center that commissions studies on societally relevant topics and broadly disseminates its research products. She heads "Thinking Like a Scientist," a national education–outreach program designed to encourage traditionally underrepresented groups (girls, people of color, and people from low-income backgrounds) to pursue science education and careers. In addition to dozens of articles and chapters on her research, Dr. Williams has authored eight books and edited three volumes. She also writes regular invited editorials for *The Chronicle of Higher Education*. Dr. Williams's research has been featured in *Nature*, *Newsweek*, *Business Week*, *Science*, *Scientific American*, *The New York Times*, *The Washington Post*, *USA Today*, *The Philadelphia Inquirer*, *The Chronicle of Higher Education*, and *Child Magazine*, among other media outlets. She served on the editorial review boards of the journals *Psychological Bulletin*, *Psychological Science in the Public Interest*, *Applied Developmental Psychology*, and *Psychology, Public Policy, and Law*. Dr. Williams is a fellow of the American Psychological Society and of four divisions of the American Psychological Association (APA). In both 1995 and 1996 her research won first-place awards from the American Educational Research Association. She received the Early Career Contribution Award from APA Division 15 (Educational Psychology); the 1997, 1999, and 2002 Mensa Awards for Excellence in Research to a Senior Investigator; and the 2001 APA Robert L. Fantz Award for an Early Career Contribution to Psychology in recognition of her contributions to research in the decade following receipt of the PhD.

I

SETTING THE STAGE

INTRODUCTION: STRIVING FOR PERSPECTIVE IN THE DEBATE ON WOMEN IN SCIENCE

WENDY M. WILLIAMS AND STEPHEN J. CECI

What is the most reliable and current knowledge about women's participation in science? Answers to this question have caused many people both within and outside of the academy to wonder: Do more women belong in science, math, and other technical fields? Is the lack of women in these fields a consequence of less ability—or simply less interest? Are there innate differences in some kinds of ability that explain the unsettling statistics, or is culture to blame? Put another way, is society holding girls and women back, or are they just not interested or intellectually equipped? To what extent are noncognitive factors responsible for the lesser representation of women—factors like institutional barriers and discrimination, demands placed on women outside of work, and unforgiving tenure clocks that collide with women's biological clocks? Also, if the number of women entering science, math, and other technical fields could be changed, should it be and what level of resources should one invest in such a change? Pondering these questions makes all of us reconsider the beliefs of our youth and the views of our parents, teachers, colleagues, and friends. New information on changing patterns of achievement and behavior has dramatically redrawn the picture, and here in Part I, "Setting the Stage," some of these issues are explored. In Part II, "Essays," well-known gender researchers offer their views on this issue, and in Part III, "Conclusion," we attempt to provide a synthesis and analysis of these views.

The essays in Part II of this book provide a context for discussing these issues within the larger social debate about the role of women in the labor force (see chap. 1, by Virginia Valian, a discussion of women's underrepresentation in leadership positions, even in professions in which they are numerically dominant, for example, nursing). We consider the meaning of "ability" in a meritocratic culture (see the concluding essay, "Are We Moving Closer and Closer Apart?"); affirmative action programs and policies (see the section in this chapter titled "The Work and Family Context"); the country's "science agenda"; and our relatively poor national performance in science internationally, a point alluded to in numerous essays herein. Why have these issues generated so much heat and so little light over so many decades? Although these larger issues are addressed in several essays, the main focus of this volume is on the role of cognitive sex differences in women's participation in science and engineering. Specifically, we examine the question of how much of the variance in successful scientific performance is attributable to cognitive differences between men and women, and we ask the following overarching question: How important are cognitive sex differences to women's participation in science and engineering? Of course, no discussion of cognitive sex differences can take place in a social vacuum. Readers will also find discussions of many noncognitive factors, such as willingness to work excessively long hours at one's science job; the demands outside of the job that impinge on women's science participation (see section "The Work and Family Context," later in this chapter); and why there continues to be debate about the meaning of the constructs *ability*, *achievement*, and *intelligence* (see the final essay of this volume, "Are We Moving Closer and Closer Apart?").

One goal of this volume is to replace readers' emotional and political biases with a broader appreciation of the issue based on solid empirical science, written accessibly for nonspecialists. Consider that Harvard University president Lawrence H. Summers ultimately unseated himself by, among other things, suggesting the possibility of innate gender differences, thereby igniting outcries on college campuses across the United States that were soon echoed throughout the media. Yet the debate has sometimes been glib, directionless, and overwhelmingly underinformed by scientific evidence. This book propels the debate forward by critically synthesizing and interpreting the top scholars' arguments. In Part II, we present a collection of 15 essays written by many of the top researchers on gender differences in cognition in the United States, Canada, and the United Kingdom, reflecting the diversity of views on the topic. As readers will note, the essayists often resummarize the same or similar data but interpret them quite differently. One of the most interesting things about reading a volume such as this is what one learns about how, precisely, multiple scientists analyze and translate the same data in different ways.

In Part II, we have gathered the leading experts, many of whom have spent decades studying precisely the issues under scrutiny in this volume, and asked them to provide the most compelling evidence to support their points of view. The reader will hear from multiple sides that take multiple perspectives; there are no glib answers offered. Armed with accurate information, readers can decide for themselves and craft informed opinions based on scientific evidence, not politics or personal beliefs. The experts provide the tools for this journey. Not everyone will appreciate hearing from the "other side," of course, and several colleagues have lobbied us to exclude certain points of view they find unacceptable. We have chosen to include all points of view—provided they are supported by empirical evidence and that they move beyond political testimonials to present scientific findings to make their case.

This volume provides a unique perspective on the causes and consequences of the dearth of women in certain fields of science. We bypass emotion, rhetoric, and politics to explore the hard science underlying gender differences in cognitive ability and their origins, along the way confronting the myriad noncognitive influences mentioned previously. We intend to leave readers with an informed understanding; we hope to replace their initial biases with biological and cultural evidence. Finally, at the end of this volume we have included a list of "Questions for Discussion and Reflection," which can be used to spur class discussions or serve as topics for written assignments.

RELEVANT HISTORY: THE SIGNIFICANCE OF SCORING IN THE EXTRAORDINARY RANGE OF ABILITY

We begin by considering recent history relevant to the debate on women in science. In 1995, Larry Hedges and Amy Nowell published an analysis of sex differences in cognitive abilities. Because of the prominence of the journal in which their article appeared (*Science*), it soon was viewed as an authoritative source on sex differences and has been cited over 140 times in other scientific articles. Hedges and Nowell examined six studies, all of which drew on national probability samples of adolescents and young adults published between 1960 and 1994. They found that male–female ability distributions differed quite a bit in these studies. The differences were especially large at the tails of the distributions, specifically, the top 1%, 5%, and 10% (as well as the bottom 1%, 5%, and 10%). Males excelled over females in science, math, spatial reasoning, and social studies as well as in various mechanical skills. Females excelled at some verbal abilities, associative memory performance, and perceptual speed. There were modest differences at the center of the distribution (the average scores for males and females were similar), but the greater variability or spread of the male scores meant that there were large asymmetries at the tails or extremes of the distribution. As just one of their dramatic findings indicates, males outnumbered females in the top 1% of mathematics and spatial reasoning ability by a ratio of 7:1. Hedges and Nowell (1995) concluded their analysis as follows:

> The sex differences in mathematics and science scores, although smaller, are of concern because ability and achievement in science and mathematics may be necessary to excel in scientific and technical occupations. Small mean differences combined with modest differences in variance can have a surprisingly large effect on the number of individuals who excel. . . . The achievement of fair representation of women in science will be much more difficult if there are only one-half to one-seventh as many women as men who excel in the relevant abilities. (p. 45)

In the aftermath of their article, there were a few criticisms, mostly concerning the implications of the findings. However, by and large their findings and interpretations were not contested; they were consistent with those of other researchers. For example, Camilla Persson Benbow (1988) reported male–female ratios among the top 0.1% of adolescents—in other words, 1 in 1,000—on the Scholastic Assessment Test—Mathematics (SAT–M; formerly known as the Scholastic Aptitude Test) of nearly 10:1. And in Julian Stanley's seminal work with 450 Baltimore 12- to 14-year-olds recommended by their science and math teachers to his gifted program, 43 boys scored higher than the highest scoring girl (Stanley, Keating, & Fox, 1974). As will be seen, in recent years the size of these gaps has shrunk appreciably, although they have not disappeared. For example, the male–female ratio among those scoring 700 or more on the SAT–M before age 13, which was 13:1 in 1983 (Benbow & Stanley, 1983), shrunk to about 4:1 by 2005. Thus, secular trends in ratios of men to women at the high end of ability are not stable.

FAST-FORWARD TO 2005: OLD DATA IGNITE NEW WARFARE

Hedges and Nowell's (1995) findings caused little stir—that is, until January 14, 2005. That was when then-Harvard University president Lawrence H. Summers, speaking to an assemblage at the National Bureau of Economic Research in Cambridge, Massachusetts, on the topic of diversifying the science and engineering workforce, commented that on aggregate, more men than women perform at the highest levels in math and science (Summers, 2005): "If you do that calculation—and I have no reason to think that it couldn't be refined in a hundred ways—you get five to one (ratio of men to women), at the high end" (¶ 4).

To be clear, Summers did not claim that women could not perform at the highest levels of math and science, only that the data indicated they were not performing at the highest level in anywhere near the proportion of men. Given that his stated "five-to-one ratio" at the highest level (the top

1%) was in line with published studies such as Hedges and Nowell's (1995), one can only surmise why his remarks caused a national stir when previously published analyses had not. Although there are many potential reasons, one is that his remark seemed to suggest that women are underrepresented in scientific, technical, engineering, and math (STEM) careers because they are cognitively deficient and lack the necessary preparation as a result of biology, socialization, career choices, or all three. Although Summers did not say this explicitly, others hearing or reading his remarks inferred it. For example, Diane F. Halpern's (chap. 9, this volume) reaction to his comments was representative of many others:

> Is the underrepresentation of women in the sciences and math caused by sex differences in cognitive abilities? Of course, the real question is not neutral—it is about a presumed deficiency in women: Are there too few women with the cognitive abilities that are needed for careers in science and math? (p. 121)

This interpretation seems justified. After all, Summers opined that as a basis for the underrepresentation of women in STEM fields, factors external to the women—such as institutional discrimination, negative stereotypes about women's ability, biased promotion practices, or early socialization experiences—were probably not as important as causes of women's STEM underrepresentation as were sex differences in ability. Having said this, Summers did acknowledge that personal needs of young female faculty (he specifically mentioned child bearing and care) that are by nature poorly aligned with institutional promotion schedules, or female-unfriendly institutional policies and even stereotypes, could account for some of the gender gap in STEM fields,[1] and he lamented the fact that Harvard University did not have a startup benefit for young faculty that included child care. Putting aside this progressive suggestion, probably the single utterance that caused the biggest stir was Summers's statement that behavioral genetic studies over the past 15 years have shown that many of the differences that were once thought to be environmental are now known to have substantial biological bases. In his words,

> Most of what we've learned from empirical psychology in the last fifteen years has been that people naturally attribute things to socialization that are in fact not attributable to socialization. We've been astounded by the

---

[1]According to a transcript of Summers's (2005) remarks at the National Bureau of Economic Research, he stated, "To what extent is there overt discrimination? Surely there is some. Much more tellingly, to what extent are there pervasive patterns of passive discrimination and stereotyping in which people like to choose people like themselves, and the people in the previous group are disproportionately white male, and so they choose people who are like themselves, who are disproportionately white male? No one who's been in a university department or who has been involved in personnel processes can deny that this kind of taste does go on, and it is something that happens, and it is something that absolutely, vigorously needs to be combated. On the other hand, I think before regarding it as pervasive, and as the dominant explanation of the patterns we observe, there are two points that should make one hesitate." (He proceeded to outline two alternative accounts.)

results of separated twins studies. The confident assertions that autism was a reflection of parental characteristics that were absolutely supported and that people knew from years of observational evidence have now been proven to be wrong. And so, the human mind has a tendency to grab to the socialization hypothesis when you can see it, and it often turns out not to be true. (Summers, 2005, ¶ 5)

Coming as it did from the gatekeeper of one of the world's great institutions of higher learning, the insinuation of biologically based sex differences in cognition, coupled with an accusation that advocates of greater equity for females in science were grasping at weak socialization explanations, was radioactive, setting off debates and protests on campuses across the nation. In his role as president of Harvard University, Summers's policies are often emulated by less prestigious institutions. He asserted that the gender imbalance at the right tail of the ability distribution was stubborn, perhaps even biologically rooted, and that regardless of its roots, it was the basis for the dearth of female scientists, engineers, and mathematicians at places such as Harvard. In minimizing rival explanations, he said it was unlikely that the gender imbalance was the result of biased hiring, tenure, and promotion practices; negative stereotypes about women's ability; or early socialization differences, and was only secondarily the result of gender differences in career interests or female-unfriendly institutional policies.

NATURE, NURTURE, OR SOME COMBINATION OF THE TWO

As readers will see, there are evidence-based grounds for a biologically based argument in a number of the essays in this volume. Consider, for example, David C. Geary's (chap. 13) suggestion that evolutionarily important behaviors such as male–male competition involve greater reliance on the ability to represent three-dimensional space geometrically and Richard J. Haier's (chap. 8) review of the gender differences in brain functioning. See also Doreen Kimura's (chap. 2) argument about the role of prenatal and postnatal hormones on spatial cognition and Simon Baron-Cohen's (chap. 12) argument that baby girls come into the world with an orientation toward people and baby boys toward objects, which leads them down different paths of interests. In addition, Ruben C. Gur and Raquel E. Gur (chap. 14) suggest that male brains are optimized for enhanced connectivity within hemispheres, whereas female brains are optimized for communication between the hemispheres, especially in language processing and posterior brain regions as indicated by the larger callosal splenia. If true, this differentiated brain morphology might be manifested in distinct gender-linked processing styles.

However, readers will also see that there are evidence-based grounds for explaining the gender differences in nonbiological and nonevolutionary terms (Jones, Braithwaite, & Healy, 2003), as Virginia Valian (chap. 1), Elizabeth S. Spelke and Ariel D. Grace (chap. 4), Nora S. Newcombe (chap. 5), Janet Shibley Hyde (chap. 10), and others argue cogently. For example, sex differences in cognitive ability within the United States are smaller than between-country differences, with females from some countries outperforming U.S. and Canadian males more often on tests of such ability than U.S. and Canadian males outperform U.S. and Canadian females. The way messages are framed about ability matters significantly, as Carol S. Dweck (chap. 3) shows, and a host of social factors, such as preferences and desire to balance work and family, all seem to be important (see data by David S. Lubinski and Camilla Persson Benbow, chap. 6), as Summers himself acknowledged.

In the wake of Summers's remarks, there was an outpouring of stories and editorials in the national media (e.g., Lally, 2005; ScienceFriday Inc., 2005). There was also a host of policy summits on the topic, including the commissioning of a yearlong report by the National Academy of Sciences' Committee on Science, Engineering, and Public Policy; a special issue of the journal *Psychological Science in the Public Interest* (forthcoming); and an open convocation titled "Maximizing the Potential of Women in Academe: Biological, Social, and Organizational Contributions to Science and Engineering Success," held at the National Academy of Sciences on December 9, 2005.

THE NEED FOR THIS VOLUME

The plan for this volume was also hatched in the aftermath of Summers's remarks. We felt that the debate surrounding his comments, at least as it played out in the print and electronic media, was often superficial and unsupported by scientific evidence. Why was this the case? We are not sure, but the quotes in the media frequently evidenced a lack of awareness and acknowledgment of the full corpus of scientific findings (or, for that matter, the full corpus of Summers's remarks). For example, one letter signed by approximately 80 eminent leaders in science, engineering, and education omitted important findings that if acknowledged would have softened their claims (Muller et al., 2005). Yet the debate over whether there are gender differences in cognition and whether these can account for the lack of women in science, math, and other technical careers is an area that has a venerable history, dating back several decades to Eleanor Maccoby and Carol Jacklin's (1974) seminal book, and includes hundreds of published scientific studies since then.

For these reasons, we invited top researchers in the debate about gender differences in cognition to explain their evidence-based positions as accessibly as possible. Note the descriptor *evidence-based*. The authors were enjoined to support all assertions and opinions with hard empirical data and give crucial references so skeptical readers could evaluate the conflicting evidence for themselves. The time for moving beyond slogans and rallying cries seems overdue. A presentation of all evidence-based views seems critically important even if it means that some positions may be offensive to some readers, as they were to some of our colleagues who read and reviewed this volume. In fact, the media have been chided roundly for failing to investigate the evidentiary basis for the myriad claims and counterclaims on this topic. In 1998, Diane Ravitch, former assistant secretary for educational research and improvement and counselor to the U.S. Department of Education, wrote an editorial in the *Wall Street Journal* in which she blasted the media (specifically a report by the American Association for University Women [1992]) for accepting claims that sex differences in scientific careers were the result of factors such as higher self-esteem among boys or teachers showering more attention and praise on boys. She argued that such ideas find little support in empirical studies:

> The schools, we were told, were heedlessly crushing girls' self-esteem while teachers (70% of them female) were showering attention on boys. Worst among their faults, according to the report . . .was that the schools discouraged girls from taking the math and science courses that they would need to compete in the future. The report unleashed a plethora of gender-equity programs in the schools and a flood of books and articles about the maltreatment of girls in classrooms and textbooks. (Ravitch, 1998, p. 1)

Ravitch went on to criticize claims made in the media by drawing on recently published data from the U.S. Department of Education showing that, far from failing girls, schools were doing a good job of closing gender gaps in mathematics and science. With the sole exception of high school physics, in which 27% of boys compared with only 22% of girls were enrolled, girls were taking as many courses in mathematics and science as boys, and this state of affairs had been true at least since 1990. For example, Ravitch pointed out that female high school graduates in 1990 had higher enrollments than boys in 1st- and 2nd-year algebra and geometry; among the graduates in 1994, there were few sex differences in precalculus, advanced placement (AP) calculus, trigonometry, statistics, and a host of science courses; and, in fact, female students were more likely to enroll in chemistry and biology than were male students. Overall, 43% of female graduates took a rigorous college-preparatory program in 1994 compared with only 35% of male graduates.

Criticisms such as Ravitch's could be leveled at many of the claims about sex differences found in today's electronic and print media, claims that support all sides of this often contentious topic. Elizabeth Spelke (2005) pointed out a series of pseudo-explanations of sex differences once popular in the media but now known to be false. It is for precisely this reason that we chose eminent scientists as authors for this volume—they all support their positions with empirical findings. References to the scientific basis of each claim are included to enable readers to go further, should they wish. This evidentiary basis enables us to come to a shared understanding of whether there are sex differences in cognitive ability and, if there are, what role they play in career imbalances.

KNOWNS AND UNKNOWNS ABOUT SEX DIFFERENCES IN COGNITIVE ABILITY

The essays in this volume show that the pattern of sex differences is much more nuanced than their depiction in the popular media (e.g., male = right brain; female = left brain). If readers have heard anything about sex differences in cognitive ability, most have probably heard that men excel at skills subserved by the right side of their cortex, such as quantitative and spatial ability, and that women excel at skills subserved by their left cerebral cortex, such as verbal ability. However, the actual differences between the sexes are far more complex.

One reason the story is complicated is because there are so many ways of looking at sex differences on cognitive measures. The conclusion will depend on the way one approaches the question. For example, one can look at differences in the following categories:

- at various ages, which can change dramatically, as will be seen;
- in various cultures, some of which show sex differences and some of which do not;
- on various tests, not all of which exhibit similar trends;
- at various points in the score distribution, not all of which show sex differences;
- in various course grades, which sometimes do not coincide with differences in test scores for those same subjects;
- in persistence within a major or profession;
- among those in the top 1% or 5% or 50% on some standardized score such as the SAT–M or the Graduate Record Examination;
- in AP credits earned in math and science;
- in rigor and number of math and science classes taken;
- in average scores on AP calculus; and
- in cohorts, as seen in Part II of this volume.

For example, Shayer, Ginsberg, and Coe's (in press) analyses demonstrate that in 1975 there was a substantial male advantage in the mean scores on the British Volume and Heaviness test but that this advantage disappeared by 2004. These changes in sex differences were most pronounced between 2000 and 2004. Relatedly, as mentioned earlier, the male advantage at the extreme right tail (those scoring 700 or more on the SAT–M before age 13), which was 13:1 in 1983 (Benbow & Stanley, 1983), shrunk to about 4:1 by 2005, showing instability in cohort differences and secular trends in ratios of males to females at the high end.

The question of whether there are sex differences depends on when and where one looks. At times, at some points in the distribution and on some dependent measures (but not others), and in some cultures but not in others, males and females are far more alike than different, whereas at other times the differences are fairly large. Thus, when speaking about the existence of sex differences in cognitive performance, the answer must be "it depends."

Because of the nature of this volume (which probes the causes for the dearth of women in science), we are less concerned about sex differences at the 50% mark of the distribution. Everyone agrees that this is not where the action is; females do as well or better than males in math and science, on average. They get better grades, take more demanding coursework, matriculate in college in greater numbers, graduate in higher numbers (57% of all graduates of 4-year programs are now female), and profess greater postsecondary professional aspirations. Scientists, especially those in mathematically intensive fields, do not come from the middle of the ability distribution. Some argue that they come from the most extreme part of the right tail, the top 0.1% of scorers in mathematics, for example (see the essay by David S. Lubinski and Camilla Persson Benbow, chap. 6).

Consider Summers's (2005) description of scientists at the top 25 research universities:

> One is not talking about people who are two standard deviations above the mean. And perhaps it's not even talking about somebody who is three standard deviations above the mean. But it's talking about people who are three and a half, four standard deviations above the mean in the 1 in 5,000, 1 in 10,000 class. (¶ 4)

Even if we disagree with his statement that scientists come from the top 0.01% (1 in 10,000), we are still talking about a dearth of women in careers whose members are derived almost entirely from at least the top 5% of the score distribution. In light of this fact, we focus on measures at the upper end of the range rather than in the middle, where women are doing as well as— or better than—men by most standards. Let us return to the complexity argument now and try to describe why the findings are so nuanced.

In contrast with typical categorical assertions about male superiority on quantitative tasks and female superiority on verbal tasks, males actually excel on some verbal tasks, and females actually excel on some quantitative tasks. Although some differences are in dispute, most are not. Here is what is generally agreed on (for additional details, see the essays by Elizabeth S. Spelke and Ariel D. Grace [chap. 4]; Diane F. Halpern [chap. 9]; Janet Shibley Hyde [chap. 10]; and David C. Geary [chap. 13]): Females tend to have average scores that are superior on tests of verbal fluency, arithmetic calculation, associative memory, perceptual speed, and memory for spatial locations. In contrast, males tend to have average scores that are somewhat superior on tests of verbal analogies, mathematical word problems, and memory for the geometric configuration of landscapes. In addition, far from the monolithic stereotype of female superiority in verbal domains and male dominance in quantitative domains, females excel at some forms of calculation and are better at spatial location memory, and males tend to excel at spatial reasoning as well as at some forms of verbal learning such as social studies and some forms of analogical reasoning. The magnitudes of the differences on most but not all of these tasks are fairly small, leading Janet Shibley Hyde and others to conclude that the sexes are actually far more alike than different.

The one skill that stands out as representing a large magnitude gap in favor of males involves mental rotation. This skill is involved in tasks for which three-dimensional objects are shown at different orientations, and one must determine whether they are the same object, or on tasks in which one is asked to judge if a two-dimensional piece of paper can be folded into a three-dimensional shape. In terms of statistical conventions, the effect size of male superiority is large on these types of tasks requiring mental rotation. There is also some suggestion that this skill is especially important for female success at higher level mathematics. For example, Casey, Nuttall, Pezaris, and Benbow (1995) found that the sex difference on the SAT–M was eliminated in several samples when the effects of mental rotation ability were statistically removed. This suggests that skill at mental rotation and related spatial cognition may affect certain mathematical abilities. It is important to note that this relation was found to be much stronger for females than males, suggesting that women may be particularly hindered by their spatial skill. However, the explanations for sex differences in mental rotation and its putative role in mathematics can be tricky, because men are likelier to form an image of one object and rotate it mentally to see if it aligns with the other object; in contrast, women are likelier to engage in a feature-by-feature comparison of the objects. Sometimes one strategy is more effective than the other, and both males and females can use both strategies—and when the sexes are constrained to use only one strategy, they tend to perform similarly. Finally, although males outnumber females at the extremes of the science and math achievement spectra, there is a great deal of inconsistency in the ratios of males to females at the high end. In some cultures, the ratios are much smaller than in others, and in some they are completely nonexistent (Feingold, 1992). Moreover, these male–female ratios appear to have changed considerably since 1983 when Camilla Persson Benbow and Julian Stanley reported a ratio of 13:1 for adolescents scoring above 700 on the SAT–M to about 4:1 today for this same age group.

In 2005, the National Center for Education Statistics issued its long-awaited report, *Trends in Educational Equity of Girls and Women: 2004.* The report, based on extensive data, also concluded that the average male and female were fairly similar:

> While females' performance in mathematics is often perceived to be lower than that of males, NAEP [National Assessment of Educational Progress] results have shown few consistent gender differences over the years, particularly among younger students. Twelfth-grade NAEP assessments in mathematics and science show no significant gender differences in achievement scores. However, females were less likely to report liking math or science. This is true despite the fact that young women take equally or more challenging mathematics and science coursework than male peers in high school (with the exception of physics, which females are slightly less likely than males to take). (National Center for Education Statistics, 2005, p. 14)

All of this demonstrates the complexity of sex differences in cognition. The differences are both more and less than what one usually thinks, and they are both more and less tractable. As Eleanor Maccoby and Carol Jacklin (1974) concluded 30 years ago in their landmark book on the psychology of sex differences, although social and emotional sex differences often show large magnitude differences in favor of males (e.g., in aggressive play, gross motor behavior, and sexual behavior), only a few cognitive skills show large sex differences. Yet are these skills the ones that are critical for successful training and subsequent careers in science and mathematics? If so, how malleable are they? How consistently are sex differences in these skills observed in other cultures? Are there known brain regions that subserve them or hormones that mediate them? For the answers to these questions and many others, see the essays in this volume by Doreen Kimura (chap. 2), Richard J. Haier (chap. 8), and Ruben C. Gur and Raquel E. Gur (chap. 14).

In view of the data showing a dearth of females scoring at the outer right tail (top 1%), one might imagine that very few women would succeed in scientific baccalaureate and graduate programs. However, this is not the case. The proportion of women earning bachelor's degrees in scientific and engineering fields has increased steadily every year since 1966. By 2001, the number of women earning degrees actually exceeded the number of men earning degrees in some scientific fields. For example, there are no longer gender differences in the number of demanding math courses taken in high school, and girls do better than boys in these courses (Gallagher & Kaufman, 2005; Xie & Shauman, 2003). Men and women earn equal grades in college math classes that are of comparable difficulty (Bridgeman & Lewis, 1996), and they major in math in nearly equal numbers. In 2000, for example, women earned 47% of bachelor's degrees in mathematics, and as Elizabeth Spelke (2005) argued, "By the most meaningful measure—the ability to master new, challenging mathematical material over extended periods of time—college men and women show equal aptitude for mathematics" (p. 955). Women are also attaining doctoral degrees in scientific and engineering fields in growing numbers: By 2001, women earned 37% of PhDs in scientific and engineering fields, up from just 8% in 1966 (National Academy of Sciences, 2005). Finally, in transnational comparisons, the sex differences in mathematics and science are sometimes nonexistent or even favor females. This suggests to some that the cause of sex differences is not biological. Otherwise, why would Singaporean or Japanese girls perform so much better than U.S. and Canadian boys, and why would some cultures have much smaller sex differences than are observed in North America?

THE SIGNIFICANCE OF THE PROPORTION OF WOMEN AT SUCCEEDING STAGES OF THE ACADEMIC CAREER

Notwithstanding the impressive progress girls and women have made in pursuing coursework and graduate degrees in science and mathematics in recent years, however, the increased representation of women among doctoral recipients has not coincided with increased faculty representation in some scientific and engineering fields. At the top 50 U.S. universities, the proportion of full professorships held by women ranges from 3% to 15%. Consider that although women earned 31.3% of chemistry PhDs between 1993 and 2003, in 2002 they were hired for only 21.5% of the assistant professorships. In the biological sciences, the drop-off between earning doctoral degrees and being hired as assistant professors was similar: Women earned 44.7% of the PhDs but were hired for only 30.2% of the assistant professorships. Similar trends exist for physics and mathematics, and similar drop-offs have been observed even when hiring of assistant professorships is based on the postdoctoral talent pool rather than the graduate school pool.

Do these drop-offs provide evidence of discriminatory hiring practices, perhaps the result of negative stereotypes about women's ability or dedication, including stereotypes some women may hold about their own ability? At first blush, this may seem a promising starting point for discussion of the basis of sex differences in some STEM fields. This is because current thinking attributes the gender gap in mathematics, at least in part, to negative stereotypes that are activated when gender is made salient in the context of an examination (Lewis, 2005). For example, women who marked the box corresponding to their sex after completing the AP calculus test scored significantly higher than their counterparts who checked off their sex box at the beginning of the exam. According to Davies and Spencer (2005), simply having students identify their sex following the AP calculus exam (rather than before it) would result in an annual increase of nearly 3,000 women eligible to begin college with advanced credit for calculus. This is presumably because directing women's attention to their gender at the start of the exam makes it salient and causes anxiety that impedes their performance.

However, there are many other factors that should be ruled out before one concludes that negative stereotypes or discriminatory hiring and promotion practices are responsible (as several of the essays in this volume point out). One obvious possibility for the declining percentages of women at successive career stages is that more women than men choose not to pursue tenure-track jobs, with their rigid and demanding expectations, and instead may pursue part-time work and adjunct posts that allow them more flexibility for family planning (Mason & Goulden, 2004). Another possibility is that having a very high aptitude for mental rotation and other forms of spatial cognition is needed for success in some STEM areas, particularly fields such as mathematics, physics, and engineering (see Richard J. Haier's essay [chap. 8, this volume]; see also Nuttal, Casey, & Pezaris, 2005). If high spatial ability mediates success in these fields, then the greater proportion of men with very high spatial and mathematical aptitude scores, rather than institutional discrimination, could be at least partly responsible for their overrepresentation. It is also possible that scoring in the top 1% or even higher confers benefits on those who earn PhDs and proceed to tenure at top research universities. For example, Wai, Lubinski, and Benbow (2005) reported that scoring in the top quartile of the top 1% (top 1 in 400) was associated with 32% receiving PhDs versus only 20% who scored in the bottom quartile of the top 1%. If there are pronounced gender asymmetries in the top quartile of the top 1%, this is relevant information.

Equally obvious is the likelihood that more than one factor is involved in producing gender asymmetries in successive stages of academic careers. As readers will see, many essayists in this volume point out that irrespective of one's view about the relative importance of nature versus nurture, no one believes that a single factor such as cognitive abilities is the sole determinant of the dearth of women scientists (e.g., see essays by Melissa Hines [chap. 7], Janet Shibley Hyde [chap. 10], Sheri A. Berenbaum & Susan Resnick [chap. 11], and Simon Baron-Cohen [chap. 12]). The situation is far too complex for any single-factor explanation. The real disagreement concerns the role of biologically based cognitive abilities in explaining the gender gap in science.

Three types of problems must be addressed by anyone arguing that women are less likely to succeed at graduate work in STEM fields as a result of their underrepresentation among the top 1% of scorers on tests of mathematics, spatial cognition, and science. The first problem with this claim is that, depending on how success is measured, women are not necessarily less likely to succeed in scientific and engineering undergraduate and graduate programs; they do as well as men in most of these (although see Doreen Kimura's essay [chap. 2] on sex differences in graduate accomplishments in mathematics and engineering), and women continue to make great strides in graduation rates.

The second problem with this assertion is that it does not accord well with transnational differences in mathematics and science. In their essays, Nora S. Newcombe (chap. 5), Diane F. Halpern (chap. 9), and Janet Shibley Hyde (chap. 10), all point out that the gap between males and females in mathematics and science is much smaller than the gap between students in the United States and students from other countries. As already mentioned, females from Japan and Singapore, for example, greatly outscore males from the United States on mathematics measures. And females in the United States have increased their representation in STEM fields steadily over the past few decades, suggesting that the gap is far from immobile and has been closing, albeit perhaps too slowly for some. However, as Richard J. Haier (chap. 8) points out in his essay, the transnational trends, although interesting, do not rule out a biological basis for gender differences in STEM fields: "Any inconsistency in itself does not argue against a possible genetic component. There are more blue-eyed people in Iceland than in Tibet, but it would be wrong to conclude from this fact that blue eye color is not genetic" (p. 116).

The third problem with the assertion that women's underrepresentation reflects insufficient cognitive ability is the statistical evidence of gender gaps in the academy. Donna Ginther (2001) analyzed the National Science Foundation's national 1973 to 1997 Survey of Doctorate Recipients data to find gender differences in salary, resources (e.g., lab space), and promotion rates—even after all observable factors such as duration on the job, publications, and grants were taken into account. The salary gap among assistant professors and associate professors was on the order of about 6% in favor of male faculty. This gap has remained fairly steady for 30 years or more. Among full professors, the salary gender gap is even larger, about 15%, and at least 6% of this gap cannot be explained in terms of observable factors.

Other, less systematic, analyses have agreed with this conclusion. In March 1999, the Massachusetts Institute of Technology (MIT) shocked the academic world by admitting that female faculty "suffer from pervasive, if unintentional discrimination" (Goldberg, 1999, p. 1). The MIT admission pinpointed the problem as it existed for senior faculty: "Many tenured women faculty feel marginalized and excluded from a significant role in their departments. Marginalization increases as women progress through their careers at MIT" (MIT, 1999, p. 2). Marginalization at MIT took the form of differences in salaries, resources, and differential treatment "despite [women having] professional accomplishments equal to those of their male colleagues" (MIT, 1999, p. 3).

However, no sooner had MIT confessed to discriminatory practices when critics came forth to argue that their data pointed to nondiscriminatory factors in sex differences in salary, lab resources, and so on. In one analysis, Patricia Hausman and James Steiger, writing in an issue of the *Independent Women's Forum*, reported results showing that senior male and female biology faculty at MIT differed, often dramatically, in publications, citations, and grant-getting, which could be the basis of differential salaries and treatment:

> We found compelling differences in productivity, influence, and grant funding between the more senior males and females that we studied. These differences may well have contributed to differences in working conditions alluded to in the M.I.T. gender study. However, few would likely question the fairness of rewarding those who publish more widely, are most frequently cited, or raise the most in grant funds. (Hausman & Steiger, 2001, p. 10)

Other studies showed that publication productivity by female biologists was lower than that of male biologists. Long (1992) showed that although women publish less early in their careers, their papers are actually cited more often than those of male biologists. Although this could be taken as evidence that female scientists do more impactful work than their male counterparts, it also can be taken to argue that there is no bias against their work—if anything, their work is given enhanced status.

These conflicting views will doubtlessly continue to stimulate discussion well into the future. Currently, the data needed to verify and extend competing arguments are not publicly available. Many of the essays in this volume relate directly to this issue. Faculty at virtually all major universities contend that they make great efforts to hire female applicants, even preferring them to male candidates when they are comparable in terms of publications and other accomplishments. Women themselves self-handicap, it is argued, by opting to follow their partners' careers at great expense to their own. Thus, women are less likely to be on tenure track and are more likely to be at small colleges in adjunct positions (see argument and data by Mason & Goulden, 2004). Men seldom sacrifice their aspirations for their partners (Williams, 2001); they are far less likely to take time off to rear families and care for elderly parents, and as Mason and Goulden (2004) showed, once a woman opts to go off tenure track or delay going on tenure track, the chances of getting on are greatly reduced. Some argue then that it is no wonder that women earn less and are less often in tenured positions at major universities. It is interesting that this same argument is used by both sides in the debate (e.g., see essays by David S. Lubinski & Camilla Persson Benbow [chap. 6] and Diane F. Halpern [chap. 9]). Finally, some argue that the existence of differences in salary and promotion rates is not in itself evidence of gender bias, because comparable variability can be found in departments that consist mainly of female faculty (e.g., nursing). We leave it to readers to judge the validity of such claims in light of evidence presented in the essays in Part II of this volume.

THE WORK AND FAMILY CONTEXT

No discussion of the dearth of women in science is complete without acknowledging the larger societal forces that impinge on women's success in the academy. There is more to success as a scientist than scoring in the right tail of the distribution of cognitive test scores, no matter how important this may prove to be. In several large-scale surveys of academic men and women, a picture emerges that is sadly inescapable: Women's success in academia is on a collision course with their success as parents and partners. According to Mason and Goulden's (2004) analysis of a nationally representative sample of doctoral recipients as well as their analysis of 4,459 tenure-track faculty working during fall and winter of 2002 to 2003 at the nine University of California campuses, the factors that affect women's success and satisfaction spill over into the family. For example, although 66% of fathers reported working over 60 hours per week at their careers, only 50% of mothers reported doing so. The reason for the lower number of hours devoted to their career is that women reported working more hours per week than men when combined across their multiple demands—career, housework, and caregiving: The totals are 101 hours per week for women with children versus 88 hours per week for men with children. (Men and women without children report working, on average, 78 hours per week across these domains.) Additionally, married mothers have reported working 4 hours less per week than do single women without children (Jacobs & Winslow, 2004, p. 117). Women with doctorates also have reported lower rates of marriage and fewer children: 41% of female academics reported being married with children versus 69% of male academics. Among academics within the first 12 years of being hired, only 30% of tenure-track women have children compared with 50% of tenure-track men. Lest one think this state of affairs is desired by academic women, the opposite appears true: Among 40- to 60-year-old academics, 40% of women expressed the wish for more children compared with only 29% of men. Finally, female academics are over twice as likely to be unmarried (28% for women vs. 11% for men) and show higher rates of divorce (144% of the men's rate). Collectively, such findings have led Mason and Goulden (2004) to conclude that

> thirty-odd years after the second-wave feminist revolution, equality in the workplace remains more of an aspiration than a reality. . . . In focusing solely on the professional outcomes as the measure of gender equality scholars have failed to acknowledge that the gap between the family outcomes of men and women, as measured by marriage, children, and divorce, is as wide as the gap in employment. (pp. 100–101)

These data are troubling for many reasons. Female academics have the deck stacked against them because they assume more of the extra-academic workload and end up working more hours in total per week than men. They also seem to be faced with a choice between children and careers in a way that is not true for men. Many proposals have been put forward to remedy this state of affairs, but all of them are problematic. It is beyond the scope of this introductory essay to analyze these proposals, but suffice it to mention that some women seem to remedy the problem at the expense of progress in the academic fields or else they create reverse-equity problems. Returning to the goal of this volume, the question that inevitably arises in discussing such troublesome data is whether women's self-reported fewer hours working at their careers may in fact be limiting their success. The answer to this question requires a type of data that we do not possess.

WHY THESE ESSAYISTS?

We did not engineer our invitations to contribute essays to Part II of this book so that there would be equal numbers of scholars on each end (or in the middle) of the culture-to-biology spectrum. Instead, we sought to invite everyone whose scholarship in this area is well known, respected, and evidence based. We made a decision not to invite essays from many articulate and thoughtful individuals whose writings appeared to us to be inspired more by ideological and aspirational factors than by robust empirical evidence. We began by consulting eminent researchers on this topic (e.g., Eleanor Maccoby, Diane F. Halpern) to solicit nominations. We vetted these nominations, ensuring that these individuals' positions were based on scientific findings and reasoning rather than personal beliefs and aspirations alone. We then persuaded the major empiricists to contribute to this volume. With only 2 exceptions out of 17, all of the top scholars whom we invited accepted our invitation.

The contributors to this volume represent a wide range of views in the debate. We believe that this inclusiveness propels the debate forward by allowing this volume to critically synthesize top scholars' arguments to enumerate shared ground that most would accept. Regardless of readers' personal answers to the questions we raised earlier about women in science, we are gratified that the field of sex-differences research has reached the state in which it is possible to have civil discourse on such a sensitive topic without one side attempting to slander the other. As Simon Baron-Cohen notes in his essay (chap. 12):

> The field of sex differences in the 1960s and 1970s was so conflict-ridden as to make an open-minded debate about any potential role of biology contributing to psychological sex differences impossible. Those who explored the role of biology—even while acknowledging the importance of culture—found themselves accused of defending an essentialism that perpetuated inequalities between the sexes, and of oppression. Not a climate in which scientists can ask questions about mechanisms in nature. (p. 159)

Regardless of readers' specific views, we hope they will find the essays in this volume riveting. All essayists defend their positions without miring readers in technical information about statistics, evolution, socialization, hormones, or other factors. Whenever technical concepts are critical to their position, essayists take pains to explain them clearly. In fact, this volume can easily be read and understood by intelligent persons in all fields and at all ages beginning in high school. Yet, the accessible writing styles of the essays does not give short shrift to the science behind the authors' views. This is excellent scholarship, in easy-to-read prose, full of provocative arguments and timely conundra. We challenge you to decide for yourselves: Why aren't more women in science?

REFERENCES

American Association of University Women. (1992). *How schools shortchange girls: The AAUW report (1992)*. Retrieved August 30, 2006, from http://www.aauw.org/research/girls_education/hssg.cfm

Benbow, C. P. (1988). Sex differences in mathematical reasoning ability in intellectually talented preadolescents: Their nature, effects, and possible causes. *Behavioral and Brain Sciences, 11*, 169–182.

Benbow, C. P., & Stanley, J. C. (1983, December 2). Sex differences in mathematical reasoning ability: More facts. *Science, 222*, 1029–1030.

Bridgeman, B., & Lewis, C. (1996). Gender differences in college mathematics grades and SAT–M scores. *Journal of Educational Measurement, 33*, 257–270.

Casey, M. B., Nuttal, R. N., Pezaris, E., & Benbow, C. P. (1995). The influence of spatial ability on gender differences in mathematics college entrance test scores across diverse samples. *Developmental Psychology, 31*, 697–705.

Davies, P. G., & Spencer, S. J. (2005). The gender gap artifact: Women's underperformance in quantitative domains through the lens of stereotype threat. In A. M. Gallagher & J. C. Kaufman (Eds.), *Gender differences in mathematics: An integrative psychological approach* (pp. 172–188). New York: Cambridge University Press.

Feingold, A. (1992). Sex differences in variability in intellectual abilities: A new look at an old controversy. *Review of Educational Research, 62*, 61–84.

Gallagher, A. M., & Kaufman, J. C. (Eds.). (2005). *Gender differences in mathematics: An integrative psychological approach*. New York: Cambridge University Press.

Ginther, D. K. (2001, February). *Does science discriminate against women? Evidence from academia, 1973–97* (Working Paper 2001–02). Atlanta, GA: Federal Reserve Bank of Atlanta.

Goldberg, C. (1999, March 23). MIT acknowledges bias against female professors. *The New York Times*, p. 1.

Hausman, P., & Steiger, J. H. (2001, February). *Confession without guilt? M.I.T. jumped the gun to avoid a sex-discrimination controversy, but shot itself in the foot*. Retrieved August 31, 2006, from http://www.iwf.org/pdf/mitfinal.pdf

Hedges, L. V., & Nowell, A. (1995, July 7). Sex differences in mental test scores, variability, and numbers of high-scoring individuals. *Science, 269*, 41–45.

Jacobs, J. A., & Winslow, S. E. (2004). Overworked faculty: Job stresses and family demands. *The Annals, 596*, 104–129.

Jones, C. M., Braithwaite, V. A., & Healy, S. D. (2003). The evolution of sex differences in spatial ability. *Behavioral Neuroscience, 117*, 403–411.

Lally, K. (2005, July 31). Aptitude aplenty. For these young women, and their mentors, science is what comes naturally. *Washington Post,* p. W08.

Lewis, D. (2005, June 24). Mathematics: Probing performance gaps. *Science, 308*, 1871–1872.

Long, J. S. (1992). Measures of sex differences in scientific productivity. *Social Forces, 71*, 159–178.

Maccoby, E. E., & Jacklin, C. N. (1974). *The psychology of sex differences.* Stanford, CA: Stanford University Press.

Mason, M. A., & Goulden, M. (2004). Marriage and baby blues: Redefining gender equity and the academy. *Annals of the American Academy of Political and Social Science, 596*, 86–103.

Massachusetts Institute of Technology. (1999, March). *A study on the status of women faculty in science at MIT.* Retrieved August 31, 2006, from http://web.mit.edu/fnl/women/women.html

Muller, C. B., Ride, S. M., Fouke, J., Whitney, T., Denton, D. D., Cantor, N., et al. (2005, February 18). Gender differences and performance in science. *Science, 307,* 1043.

National Academy of Sciences. (2005, December 9). *Convocation on maximizing the potential of women in academe: Biological, social, and organizational contributions to science and engineering success.* Washington, DC: Committee on Women in Academic Science and Engineering.

National Center for Education Statistics. (2005). *Trends in educational equity of girls and women: 2004.* Washington, DC: Author.

Nuttal, R. L., Casey, M. B., & Pezaris, E. (2005). Spatial ability as a mediator of gender differences on mathematics tests. In A. M. Gallagher & J. C. Kaufman (Eds.), *Gender differences in mathematics: An integrative psychological approach* (pp. 121–142). New York: Cambridge University Press.

Ravitch, D. (1998, December 17). Girls are beneficiaries of gender gap. *Wall Street Journal (Education),* p. 1.

Shayer, M., Ginsberg, D., & Coe, R. (in press). 30 years on—a large anti-Flynn effect? The Piagetian test volume and heaviness norms 1975–2003. *British Journal of Educational Psychology.*

ScienceFriday Inc. (2005, December 2). Gender differences in learning and recognition [Radio series broadcast]. In K. Vergoth (Producer), *ScienceFriday.* New York: Author.

Spelke, E. S. (2005). Sex differences in intrinsic aptitude for mathematics and science? A critical review. *American Psychologist, 60,* 950–958.

Stanley, J. C., Keating, D. P., & Fox, L. H. (1974). *Mathematical talent: Discovery, description, and development.* Baltimore: Johns Hopkins University Press.

Summers, L. H. (2005, January 14). *Remarks at NBER conference on diversifying the science and engineering workforce.* Retrieved July 25, 2006, from http://www.president.harvard.edu/speeches/2005/nber.html

Wai, J., Lubinski, D. S., & Benbow, C. P. (2005). Creativity and occupational accomplishments among intellectually precocious youths: An age 13 to age 33 longitudinal study. *Journal of Educational Psychology, 97,* 484–492.

Williams, W. M. (2001, July 20). Women in academe and the men who derail them: How ineffective mentorship derails women's academic careers. *Chronicle of Higher Education,* p. B20.

Xie, Y., & Shauman, K. (2003). *Women in science: Career processes and outcomes.* Cambridge, MA: Harvard University Press.

II

ESSAYS

1

WOMEN AT THE TOP IN SCIENCE— AND ELSEWHERE

VIRGINIA VALIAN

Why are there so few women in science, especially at the top? Hold on a minute. Is that the right question? That phrasing implies that science is different from other fields. Yet is it?

In one way, science does differ from other fields. A smaller percentage of women get advanced degrees in most of the natural sciences (although not biology) than in most of the social sciences, the humanities, medicine, law, business, or nursing. Yet in another way, science is the same as other professions: Women make less money and advance through the ranks more slowly not just in the natural sciences but in every field (e.g., see Valian, 1998, 2005a; for my gender tutorials, see Valian, 2005b), including nursing (Robinson & Mee, 2004).

The ubiquity of women's underrepresentation at the top provides important information about where to look to understand women's underrepresentation in science. There is a need both to look below the sur-

I thank Martin Chodorow, Howard Georgi, Cathy Kessel, and Mary Potter for comments, helpful suggestions, and references. This work was supported in part by an award from the National Science Foundation (SBE 0123609). Correspondence may be addressed to gender.tutorial@hunter.cuny.edu.

face of any particular field to understand how people are evaluated in professional settings and to understand which features of organizations give men more opportunities to be successful. There are two questions to answer. Why are there so few women at the top, even in fields like nursing and restaurant cooking, and why are there fewer women in most of the natural sciences than in other fields?

I provide the same explanation for both problems—a combination of gender schemas and the accumulation of advantage. Let us get two other possible reasons out of the way first.

ARE WOMEN LESS TALENTED THAN MEN?

One explanation for the lack of women at the top is that women are less talented than men, especially in the natural sciences and math. Three questionable assumptions appear to underlie that statement. One is that there is a single talent that determines success in the natural sciences and math; another is that existing standardized quantitative tests measure that single talent, and the third is that talents and abilities are fixed rather than malleable. All three assumptions—if accepted by young people or their teachers—are likely to reduce the number and range of people who will do creative and substantive work in the sciences. Only those with the highest standardized test scores and the most confidence would continue.

The physicist Howard Georgi, in a letter in January 2005 to Harvard undergraduates, noted that he had observed thousands of undergraduate physics students and went on to say the following:

> 1—Talent is not a unitary thing. It is multidimensional and difficult to measure or quantify precisely.
> 2—Many different kinds of talents are critical to the advancement of physics or any other science interesting enough to be worth doing.
> 3—The spread of talents within any group, sex, race, etc., is very large compared to any small average differences that may exist between such groups.
> 4—Talent can be developed and enhanced by education, encouragement, self-confidence, and hard work.
> For these reasons, I think that it is not particularly useful to talk about innate differences to explain the differences in representation of various groups in physics. Instead, I conclude that we need to try harder to teach our wonderful subject in a way that nourishes as many different skills as possible. (Georgi, 2005, ¶ 4–¶ 8)

Georgi's views mark him as an *incremental* theorist. Dweck and her colleagues (see Levy, Plaks, Hong, Chiu, & Dweck, 2001) have distinguished between people who see a given trait as fixed and unchanging—*entity* theorists—and those who see the trait as malleable and capable of increasing—

incremental theorists. Most of us are entity theorists about some traits and incremental theorists about others. For any given trait, about 85% of people fit neatly into the entity or incremental category, and that 85% is roughly equally divided between the two categories (Levy et al., 2001). If someone is an entity theorist about math and science skills and abilities, that person will treat those skills as stable and largely unresponsive to training and effort. In contrast, an incrementalist will see the skills as traits that can be developed and that can differ from one context to another. Thus, Georgi is an incremental theorist about physics.

There are consequences of holding an entity or incremental view (Levy et al., 2001). Entity theorists see groups as internally more homogeneous than incremental theorists do. For example, an entity theorist would see physicists as a more homogeneous group than an incremental theorist would. Enrico Fermi, who won the Nobel Prize in physics in 1938, was reportedly asked whether the prize winners in physics had any characteristics in common. After some thought, he replied that he could not think of a single one, including intelligence (Shucking, 1994).[1] Fermi, then, was not an entity theorist about physics. Entity theorists also see the differences between groups as larger than incremental theorists see them. Finally, entity theorists are more prone than incremental theorists to see the traits of a group, such as scientists, as being due to innate characteristics; incremental theorists allow more room for experiences and environment (Levy et al., 2001).

Which group is right about "talent" in math and physics? The U.S. educational system as a whole acts like an entity theorist about math, whereas Japan's educational system acts like an incrementalist. Japanese educators see math as a set of skills that can be taught well or badly and that require effort on a student's part. An international report of the 2002–2003 Trends in International Mathematics and Science Study compare how eighth graders in different countries perform on math tests. The overall average score (combining the domains of knowing, applying, and reasoning, with reasoning as the most sophisticated form of performance) of Japanese girls was 569 and that of Japanese boys was 571 (Mullis, Martin, & Foy, 2005). In contrast, the comparable figures for the United States were 502 for girls and 507 for boys. In both countries there is a small sex difference favoring boys—2–5 points. It is important to note, though, that the Japanese girls outperform U.S. boys by 62 points. Japan has been outperforming the United States since international testing began in the mid-1960s. Similarly, girls and boys from Singapore, with average scores of 611 and 601, respectively, performed even better—a full standard deviation better than American children. The cross-national differences dwarf the sex differences. If high test scores were the main determinant of mathematical discovery, Asians—male and female—

---

[1] I am grateful to Dudley Herschbach, cowinner in 1986 of the Nobel Prize in chemistry, for alerting me to this quotation (personal communication, July 8, 2005).

would dominate mathematics (to the best of my knowledge, however, they do not).

In the case of the most sophisticated performance—reasoning—there were, again, very small cross-sex differences and very large cross-national differences. Further, when sex differences existed, they favored girls (Mullis et al., 2005). That is contrary to earlier research, which had suggested that girls did as well as boys at standard problems but lagged behind boys with problems that required unconventional solutions. The cross-national differences were stunning. A score of 446 placed a Japanese student at only the 5th percentile in the domain of mathematical reasoning in his country; a similar score, 448, placed a U.S. student at the 25th percentile. A Singaporean student at the 75th percentile in her country, with a score of 645, would be slightly above a child at the 95th percentile in the United States, with a score of 638. Mathematical competence can be nurtured.

Let us look now at what test scores tell us about who gets advanced degrees in science. In the United States (and most but not all other nations where scores are available), girls consistently have less variable distributions than boys and thus are less likely than boys to score at the top end in quantitative tests (for summary data, see Hedges & Nowell, 1995; Lubinski & Benbow, 1992). It is not known why that is the case (for reviews, see Halpern & LaMay, 2000; Valian, 1998), but we do know that the difference at the top end is decreasing, supporting an incrementalist perspective. In 1983, for example, seventh- and eighth-graders in a national sample of gifted children showed a large sex difference between children who scored 700 or above on the math SAT. There were 13 boys for every girl (Benbow & Stanley, 1983). By 2005 that difference had plummeted to 3 to 1 (Brody & Mills, 2005) or 4:1 (Benbow, personal communication, July 6, 2006). Such a striking change is incompatible with the idea of a fixed difference between boys and girls.

In addition, the differences do not predict U.S. youngsters' intentions to major in science in college (Xie & Shauman, 2003). Neither average sex differences on standardized math and science achievement tests taken in the 8th, 10th, and 12th grades nor sex differences among the top 5% of test takers account for the sex differences in students' intended college majors (Xie & Shauman, 2003). Twice as many boys as girls intend to go into science, but that sex difference is not explained by sex differences in test scores. If anything, taking scores into account exacerbates the sex difference. Top-scoring girls are a particular casualty of science and math education in the United States—not a desirable result; we do not succeed in nurturing girls' talent.

A different kind of measure focuses on the educational outcomes for young adolescents who demonstrate early high performance on the quantitative SAT (Benbow, Lubinski, Shea, & Eftekhari-Sanjani, 2000). Two groups of 12- to 14-year-olds who scored in the top 1% of their age group in the 1970s were surveyed 20 years later at age 33; in the initial group, boys outnumbered girls by a little more than 2:1. Let us look at how the percentages of males and females who achieved undergraduate degrees in various fields compared with the percentages who achieved PhDs in those same fields. The natural sciences and math lost large numbers of high-scoring males and an even larger proportion of high-scoring females. The higher education system in the United States wastes female talent.

Take mathematics as an example. The same percentage (about 10%) of high-scoring boys and girls received undergraduate degrees in math. Because there were a little more than twice as many boys as girls among the high scorers to begin with, about 42 boys compared with 18 girls earned a bachelor's degree in math. At the PhD level, the comparison is considerably worse: About 9 boys and 1 girl ended up with a PhD in math. An initial 2.3:1 difference in BAs becomes a 9:1 difference in PhDs (calculated from Benbow et al., 2000, Table 1, Cohort 1). High-scoring girls are not being retained in math and science at the same rate as high-scoring boys. Thus, whether we look at population data (Xie & Shauman, 2003) or top scorers (Benbow et al., 2000), we see the same picture. The U.S. education system does not retain girls at the same rate as boys. Furthermore, there is greater attrition of both boys and girls in math and the natural sciences than in other fields.

In summary, test performance cannot explain the low representation of women in math and natural science.

Are Women Less Interested in Professional Careers Than Men Are?

Another reason for women's underrepresentation in the sciences may be that women are less interested in a professional career—in science or any other field—than men are. Again, there are unspoken assumptions in the phrasing of the hypothesis. One is that people make their choices in an unconstrained manner and are unaffected by the support and encouragement they receive or fail to receive. Another is that having a high-powered professional life and having a rich personal life are incompatible; you have to pick one or the other. A third is that anyone who is talented and works hard will be successful, regardless of their sex, race, age, and so on.

Because people are affected by encouragement, support, and expectations (see Valian, 1998, for a review of relevant experiments), it is difficult to evaluate people's choices. What would men do if they, like women, were expected and encouraged to take care of children? What would women do if they, like men, were expected and encouraged to have a professional life?

One reason to question the sufficiency of an explanation that emphasizes women's interest in a family is that women pay a price in rate of advancement for being women, even if they do not have children. Women without children do not advance as fast as men (for academic science, see Long, 2001; for law, see Wood, Corcoran, & Courant, 1993). Thus, although fathers do less and mothers do more than their fair share of child care, that alone does not account for the fact that women without children are less successful than men. There is a professional cost to women and to society of women's having children—absence from full-time employment (Long, 2001; Xie & Shauman, 2003). But women with children who remain as full-time academics publish the same amount as women without children (Long, 2001; Valian, 1998). They also have careers that are very similar to those of childless women. In summary, women's interest in a full life cannot explain their sparse numbers in science or at the top of different professions.

GENDER SCHEMAS AND ACCUMULATION OF ADVANTAGE

Why, then, are women underrepresented in science and underrepresented at the top in all professions? This is where gender schemas and the accumulation of advantage come in. *Schemas* are hypotheses that are used to interpret social events (Fiske & Taylor, 1991). Schemas are similar to stereotypes, but the term *schema* is more inclusive, more neutral, and more appropriate because it brings out the protoscientific nature of social hypotheses. Social schemas are necessary. One cannot treat every person one meets as if the social group to which they belong is irrelevant; it often is relevant and provides valuable information. But schemas—being schematic—oversimplify and thus can lead to mistakes.

Gender schemas are hypotheses about what it means to be male or female, hypotheses that all people share, male and female alike. Schemas assign different psychological traits to males and females (Martin & Halverson, 1987; Spence & Helmreich, 1978; Spence & Sawin, 1985). As folk psychologists, we think of males as capable of independent action, as oriented to the task at hand, and as doing things for a reason. We see females as nurturant, expressive, and behaving communally. In brief, men act; women feel and express their feelings. And our beliefs have support. In questionnaires, men endorse more "instrumental" characteristics and women endorse more "expressive" characteristics. The sexes overlap, as they do on every measure of behavior, perception, and cognition; but there are broad differences.

The main answer to the question of why there are not more women at the top is that gender schemas skew our perceptions and evaluations of men and women, causing us to overrate men and underrate women. Gender schemas affect judgments of people's competence, ability, and worth.

Consider two experiments on judgments of women's competence. Both were published in 2004 and demonstrate some of the effects of gender schemas. The first investigated how males and females rated people who were described as being "assistant vice presidents" in an aircraft company (Heilman, Wallen, Fuchs, & Tamkins, 2004). The evaluators read background information about each person, the job, and the company. In half of the cases, the person was described as about to have a performance review; thus, evaluators did not know how well the person was doing in the job. In the other half of the cases, the person was described as having been a stellar performer. The evaluators' job was to rate how competent the employees were and how likeable they were.

When no information was given about how well people were doing in the job, evaluators rated the man as more competent than the woman and rated them as equally likeable. When the background information made clear that the woman was extremely competent, however, the ratings changed. Evaluators now rated the man and the woman as equally competent, but they rated the woman as much less likeable than the man. They also perceived the woman as considerably more hostile than the man.

Thus, in evaluating a woman in a male-dominated field, observers saw her as less competent than a similarly described man unless there was clear information that she was competent. In that case, they saw her as less likeable than a comparable man. Notably, as is the case in almost all such experiments, there were no differences between male and female raters. Both male and female raters saw competence as the norm for men and as something that has to be demonstrated unequivocally for women. Both male and female raters saw competent men as likeable. Neither male nor female raters saw competent women in male-dominated positions as likeable.

Does likeability matter? In a follow-up experiment, the experimenters described targets as high or low in competence and as high or low in likeability. People rated the targets who were high in likeability as better candidates for being placed on a fast track and as better candidates for a highly prestigious upper level position. One cannot tell women just to be competent, because likeability can make the difference in whether people get rewards. Again, there were no differences between male and female raters.

The second study demonstrated how people shift their standards to justify a choice that seems a priori reasonable to them (Norton, Vandello, & Darley, 2004). In this experiment, gender schemas determined what seemed reasonable. The experiments asked male undergraduates to select a candidate for a job that required both a strong engineering background and experience in the construction industry. The evaluators rated five people, only two of whose resumes were competitive. One candidate had more education—both an engineering degree and certification from a concrete masonry association—than the other, who only had an engineering degree. The other candidate had more experience—9 years—than the first, who only had 5 years.

In the control condition, the candidates were identified only by initials. Here, the evaluators chose the candidate with more education three fourths of the time, and education was most often cited as the most important determinant of their decision. In one of the experimental conditions, a male name was given to the resume that had more education and a female name to the resume that had more experience. Here, too, evaluators chose the candidate with more education three fourths of the time and also rated education as very important. In the second experimental condition, a female name was given to the resume with more education and a male name to the resume with more experience. In this condition, less than half the evaluators picked the person with more education and less than a quarter said that education was the most important characteristic.

Men look more appropriate than women for the job of construction engineer, whether they have more education or more experience. The standards by which people judge others will shift depending on a priori judgments about their goodness of fit. Gender schemas help determine goodness of fit. Shifting standards are in operation at work, when job candidates are evaluated, and at home, when heterosexual couples make decisions about whose profession is more important. Thus, even though most people are genuinely meritocratic and egalitarian, and even though people's estimates of sex differences correlate well with psychological measures of sex differences (Swim, 1994), people's implicit evaluations of performances are in tune with gender schemas.

These experiments and others like them are relevant to women's underrepresentation in most natural science fields and underrepresentation at the top in every field. To take the second issue first, the experimental data show that both men and women slightly overrate men and underrate women in professional domains. Women appear to both men and women to be less competent. The small imbalances in evaluation and perception add up to advantage men and disadvantage women. It is like interest on an investment. If X has a slightly higher interest rate than Y, X will end up with more money than Y down the line, thanks to the "miracle" of compound interest.

A similar "miracle" happens in the professional world. Success is largely the accumulation of advantage, parlaying small gains into bigger ones (Merton, 1968). If you do not receive your fair share of small gains because of the social group you belong to, you—and your group—will be at a disadvantage. The miracle of compound interest means that someone who receives an extra quarter percent of interest than you will be in better shape than you 10 years later.

A computer simulation (Martell, Lane, & Emrich, 1996) showed the importance of even tiny amounts of bias. The researchers simulated an eight-level hierarchical institution with a pyramidal structure. They staffed this hypothetical institution with equal numbers of men and women at each level. The model assumed a tiny bias in favor of promoting men, a bias accounting for only 1% of the variability in promotion. After many series, the top level was 65% male. Even very small amounts of disadvantage accumulated.

Thus, even in a work environment in which everyone intends to be fair—and believes they are being fair—men are likely to receive advantages in evaluations that women do not. Over time, those advantages mount up, so that men reach the top faster and in greater numbers than women do. Each individual event in which a woman does not get her due—is not listened to, is not invited to give a presentation, is not credited with an idea—is a mole-hill. Well-meaning observers may tell the woman not to make a mountain out of a molehill. What they do not understand is what the notion of the accumulation of advantage encapsulates. Mountains are molehills, piled one on top of the other.

Because gender schemas and the accumulation of advantage operate in all the professions, men in general will have an easier time than women getting to the top in all the professions. The role of gender schemas can be extended to the natural sciences at all levels. The schema of a natural scientist is more compatible with the schema for men than the schema for women. Consider again the specifics of the schemas. People see males as capable of independent action, as oriented to the task at hand, and as doing things for a reason. They see females as nurturant, expressive, and behaving communally. The natural sciences seem particularly compatible with the schema for males and out of keeping with the schema for females.

Where do gender schemas come from? I have proposed that, like race schemas (Hirschfeld, 1997), they are cognitive in origin rather than motivational or emotional (Valian, 1998). Humans create categories, and the ability to do that is an important step in scientific reasoning. In addition, I propose, humans aim first for binary categories that have nonoverlapping characteristics. We are of course capable of creating more categories and of noticing overlap, but on grounds of simplicity, we prefer to create two categories and to have those two be as distinct as possible. A two-category system is fast and efficient, even if it sometimes leads to error. In the case of sex, I suggest that the two categories are based on observation of a qualitative physical difference between the sexes. Females give birth and physically nurture their young. People draw an analogy from the physical to the mental and see females not only as physically nurturing but as metaphorically nurturing. The next step is to add other traits that seem highly compatible with nurturance to the schema for females and to construct an "alternative" set of characteristics for men (Parsons & Bales, 1955). Although the reasoning behind gender and race schemas makes sense, schemas lead to errors in evaluation. Worse, they can lead to the development of behaviors and traits that appear to show the validity of the initial distinction.

Many false beliefs are eventually dispelled by education. A flat earth looks like a natural hypothesis, but education provides data and theory to show that the earth is round. There is a limit to how far astray one can go with a belief about a flat earth even if one keeps it forever; one can never make the earth flat. In the psychological domain, however, one can make the data fit the theory. One can discourage females from high professional achievement and discourage males from nurturing parenthood. One can create limits where none intrinsically exist. The aim of this essay is to present the alternative—to use scientific experiments and reasoning to erase arbitrary limits.

REFERENCES

Benbow, C. P., Lubinski, D. S., Shea, D. L., & Eftekhari-Sanjani, H. (2000). Sex differences in mathematical reasoning ability at age 13: Their status 20 years later. *Psychological Science, 11*, 474–480.

Benbow, C. P., & Stanley, J. C. (1983, December 2). Sex differences in mathematical reasoning ability: More facts. *Science, 222*, 1029–1030.

Brody, L. E., & Mills, C. J. (2005). Talent search research: What have we learned? *High Ability Studies, 16*, 97–111.

Fiske, S. T., & Taylor, S. E. (1991). *Social cognition* (2nd ed.). New York: McGraw-Hill.

Georgi, H. (2005, January 21). *Talent, skills in math and science hard to quantify.* Retrieved November 26, 2005, from http://www.thecrimson.com/article.aspx?ref=505377

Halpern, D. F., & LaMay, M. L. (2000). The smarter sex: A critical review of sex differences in intelligence. *Educational Psychology Review, 12*, 229–246.

Hedges, L. V., & Nowell, A. (1995, July 7). Sex differences in mental test scores, variability, and numbers of high-scoring individuals. *Science, 269*, 41–45.

Heilman, M. E., Wallen, A. S., Fuchs, D., & Tamkins, M. M. (2004). Penalties for success: Reactions to women who succeed at male gender-typed tasks. *Journal of Applied Psychology, 89*, 416–427.

Hirschfeld, L. (1997). The conceptual politics of race: Lessons from our children. *Ethos, 25*, 63–92.

Levy, S. R., Plaks, J. E., Hong, Y., Chiu, C., & Dweck, C. S. (2001). Static versus dynamic theories and the perception of groups: Different routes to different destinations. *Personality and Social Psychology Review, 5*, 156–168.

Long, J. S. (Ed.). (2001). *From scarcity to visibility: Gender differences in the careers of doctoral scientists and engineers.* Washington, DC: National Academy Press.

Lubinski, D. S., & Benbow, C. P. (1992). Gender differences in abilities and preferences among the gifted: Implications for the math–science pipeline. *Current Directions in Psychological Science, 1*, 61–66.

Martell, R. F., Lane, D. M., & Emrich, C. (1996). Male–female differences: A computer simulation. *American Psychologist, 51*, 157–158.

Martin, C. L., & Halverson, C. (1987). The roles of cognition in sex role acquisition. In D. B. Carter (Ed.), *Current conceptions of sex roles and sex typing: Theory and research* (pp. 123–137). New York: Praeger.

Merton, R. K. (1968, January 15). The Matthew Effect in science. *Science, 159*, 56–63.

Mullis, I. V. S., Martin, M. O., & Foy, P. (2005). *TIMSS 2003 international report on achievement in the mathematics cognitive domains: Findings from a developmental project.* Retrieved November 24, 2005, from http://timss.bc.edu/PDF/t03_download/T03MCOGDRPT.pdf

Norton, M. I., Vandello, J. A., & Darley, J. M. (2004). Casuistry and social category bias. *Journal of Personality and Social Psychology, 87*, 817–831.

Parsons, T., & Bales, R. (Eds.). (1955). *Family, socialization, and interaction process.* New York: Free Press.

Robinson, E. S., & Mee, C. L. (2004). Salary survey. *Nursing, 34*, 36–39.

Shucking, E. L. (1994). Review of Einstein lived here. *Physics Today, 47*, 70–71.

Spence, J. T., & Helmreich, R. L. (1978). *Masculinity and femininity: Their psychological dimensions, correlates, and antecedents.* Austin: University of Texas Press.

Spence, J. T., & Sawin, L. L. (1985). Images of masculinity and femininity: A reconceptualization. In V. E. O'Leary, R. K. Unger, & B. S. Wallston (Eds.), *Women, gender, and social psychology* (pp. 35–66). Hillsdale, NJ: Erlbaum.

Swim, J. K. (1994). Perceived versus meta-analytic effect sizes: An assessment of the accuracy of gender stereotypes. *Journal of Personality and Social Psychology, 66*, 21–36.

Valian, V. (1998). *Why so slow? The advancement of women.* Cambridge, MA: MIT Press.

Valian, V. (2005a). *Sex disparities in advancement and income.* Retrieved November 24, 2005, from http://www.hunter.cuny.edu/genderequity/equityMaterials/numbers.pdf

Valian, V. (2005b). *Tutorials for change: Gender schemas and science careers.* Retrieved November 24, 2005, from http://www.hunter.cuny.edu/gendertutorial/

Wood, R., Corcoran, M., & Courant, P. (1993). Pay differences among the highly paid: The male–female earnings gap in lawyers' salaries. *Journal of Labor Economics, 11*, 417–441.

Xie, Y., & Shauman, K. (2003). *Women in science: Career processes and outcomes.* Cambridge, MA: Harvard University Press.

2

"UNDERREPRESENTATION" OR MISREPRESENTATION?

DOREEN KIMURA

The word *underrepresentation* as the theme of this volume says much about the bias we contend with in attempting a rational discussion of sex differences. It has become standard form to assume that if there are fewer than 50% women in any cohort, the situation is undesirable and indicates some form of systemic or deliberate discrimination. We don't hear of underrepresentation of men in nursing or education, yet this would be an analogous, and equally fallacious, description. Most people assume that the lower numbers of men in these fields reflect a lesser talent or interest, and that is almost certainly correct.

Coupled with this biased view is another that sustains it: that there are no substantial differences between men's and women's cognitive profiles that cannot readily be altered by appropriate socialization. This view is so strongly ingrained in most social scientists (and in nervous politicians) that it has become a rule that socialization interpretations must be given priority over others. Such a position is basically incompatible with scientific principles, because it encourages the ignoring of a large body of opposing research.

This essay concerns sex differences in specific cognitive abilities rather than general intelligence. Nevertheless, the possibility that men have a slight advantage in overall intelligence, or IQ, is under serious discussion (Irwing & Lynn, 2005).

Cognitive differences between men and women vary greatly in magnitude across tasks as measured by effect size[1] (Halpern, 1992; Kimura, 1992) and therefore also in replicability from study to study, particularly when the number of people tested in a sample is small. Large effect sizes favoring males are found for certain spatial tasks, especially mental rotation tasks (requiring correction for object orientation), mechanical reasoning (imagining real-world interactions of mechanical items), and throwing accuracy. Large differences favoring females are found on tasks such as verbal memory (recall of words from either a list or a meaningful prose passage) and object location memory (ability to recall the locations of multiple specific objects). Smaller average differences favoring males are found on mathematical reasoning tests and spatial visualization (ability to imagine figural manipulation or rearrangement), and favoring females on verbal fluency (generating as many words as possible with a specified constraint, e.g., beginning letter) and perceptual speed (scanning for targets or matches). Many other tests also show sex differences.

However, in addition to noting average differences, we need also to look at the relative representation of men and women at the high end of any distribution of scores. Scores at the low end may be of interest for other purposes, but they are of lesser validity for predicting choice of specific occupation or profession. For example, men generally greatly outnumber women at the high end of math reasoning tests, and this difference becomes more marked as the tests become more demanding. I return to this point later.

Research into cognitive sex differences over the past half-century has shown that many human cognitive sex differences are

- significantly influenced by both prenatal and current levels of sex hormones (see Kimura, 2000, for this and following points); prenatal androgen levels are almost certainly a major factor in the level of adult spatial ability; however, even in adulthood, variations in hormone levels (across the menstrual cycle in women and across seasons and time of day in men) are associated with variations in specific cognitive abilities;
- present very early in life, before major differences in life experience (e.g., Levine, Huttenlocher, Taylor, & Langrock, 1999); thus not all cognitive sex differences develop gradually through postelementary school years; of course, even those that do not appear until after puberty are not necessarily determined solely

---

[1]Effect size refers to the magnitude of the difference between two means, taking into account the variability or dispersal of the scores around each mean. If variability is small, there is less overlap between the groups, so the effect size is larger. If variability is large, effect size will be smaller.

by experience but may be influenced by the pubertal alterations in sex hormone levels;

- present across cultures that vary in social pressures to conform to a gender norm; this has been documented for both math reasoning and spatial ability (e.g., Geary & DeSoto, 2001);
- apparently uninfluenced by systematic training in adulthood; although both sexes benefit by short-term intensive training on spatial tasks, men's and women's scores do not converge (Baenninger & Newcombe, 1989);
- not radically changed in magnitude over the past 3 or 4 decades, a period in which women's roles and access to higher education have changed substantially (Feingold, 1996; Kimura, 2002b); and
- parallel to certain sex differences found in nonhumans in which social influences are, either naturally or by virtue of a laboratory environment, absent or minimal; for example, male rats are superior to female rats in learning spatial mazes, and these sex differences can be reversed by hormonal manipulation in early postnatal life (Williams & Meck, 1993).

The more difficult question is to what degree such differences, reliable and large as some may be, account for the differential representation of women (or men) in any field of endeavor. Common sense would dictate that the differences should influence choice of career, but the relevant research is largely correlational, which, in the absence of other converging evidence, may be open to conflicting interpretations. For example, there have been studies showing that adults who engage in spatially demanding activities also generally engaged in such activities in childhood, that is, that the two are related (Newcombe, Bandura, & Taylor, 1983). The usual, and problematic, interpretation by most social scientists is that the childhood experience determined the adult pattern. However, this may be quite wrong. It is equally possible, in fact given other evidence, probable, that those people who are natively better endowed spatially are likely to engage in such activities early in life as well as later. That is, the superior abilities influence the activities rather than the other way round. Inferences about direction of causality in correlational data are not simple, and in fact the relation between two characteristics may often be due to third factors not measured, in this instance quite probably the level of sex hormones. The hormonal studies suggest that an optimal level of prenatal androgens ("male" sex hormones) is a possibly sufficient condition for some instances of superior spatial ability (Hampson, Rovet, & Altmann, 1998; Resnick, Berenbaum, Gottesman, & Bouchard, 1986).

To return to the factors that influence the lower representation of women in science, and particularly the physical sciences, we have to consider the role of mathematical aptitude. We first must distinguish between aptitude and achievement. Throughout the school years, girls generally get better marks on most subjects, when achievement is tested, and often this has also been the case for mathematical subjects. However, when these same groups of females and males are tested on math aptitude, which may ask for solutions to problems not already rehearsed in the classroom, males get better scores (Felson & Trudeau, 1991; Wainer & Steinberg, 1992).

The scores for Scholastic Aptitude Test—Mathematics have consistently over several decades been higher for high school boys. The participants selected for Benbow's (1988) studies of mathematically precocious youth (SMPY) have consistently included a greater number of boys. Even within this select group of boys and girls, the average scores of boys are higher. The ratio of boys to girls at the high end of the distribution of scores is about 10:1. In the Putnam competition, open to all undergraduates in North America, what data we have suggest a huge preponderance of males who get the highest scores, even correcting for the larger numbers of male applicants. To date, all the recipients of the Fields medal, a prestigious award in mathematics, have been men.

Even those young women who select themselves for math interest in the SMPY study, and who do well on mathematical aptitude tests, much more often than young men prefer to enter fields of study that do not emphasize mathematics (Lubinski & Benbow, 1992). This suggests that there is a different distribution of not only talent but also interest in math-oriented activities across the sexes. That would suggest that women might have particularly low representation in those fields of science that require a high degree of math talent, such as engineering and the physical sciences.

We also know that one type of spatial ability, mental rotation, significantly predicts how well people perform on navigational tasks. These vary from strictly laboratory-type tasks (e.g., ease in learning a paper-and-pencil route or a "virtual maze" on a computer) to those that more nearly approximate real life (learning a route through a campus; Saucier et al., 2002). Men perform better than women, on average, on both the mental rotation test (which does not require navigation) and the navigational tasks, and the scores on the two types of tasks are highly related in both sexes. To that degree, mental rotation ability might well account for much of the higher representation of men in occupations such as airline pilot, for example. Shea, Lubinski, and Benbow (2001) also suggested that a combined measure of spatial visualization and mechanical reasoning ability adds substantially to accurate predictions of occupational choice, high scorers favoring engineering and architecture over law or medicine.

In contrast, one might argue, although not directly pertinent to the question of representation of the sexes in science, that the greater representation of women in secretarial work is related to their advantage in two fields. One is superior finger dexterity and the other is superior verbal memory, both quite reliable findings. Young girls show greater aptitude on both these abilities at young ages (Denckla, 1974; Ingram, 1975; Kramer, Delis, Kaplan, & O'Donnell, 1997; McGuinness, Olson, & Chapman, 1990). Of course social tradition may contribute to this situation, but one must remember that traditions have to arise somehow, and differential abilities are likely to be a major contributing factor.

Although enrollment of women in the sciences at the undergraduate level in North America has increased substantially in recent years, their representation still decreases at successive graduate levels. Moreover, research productivity in enrolled graduate students shows a male advantage in math and engineering fields, but not in education, humanities, or social sciences (Nettles & Millett, 2006). If we look at earned doctorates, Canadian data since 1998 show that men outnumber women in all fields except education, in which well over 60% of doctoral degrees go to women. In mathematical, physical, and computer sciences, men overwhelmingly receive the bulk of doctorates, with women ranging from 20% to 23%. In engineering, the picture is even more uneven, with women receiving barely 15% of doctorates. However, in biological sciences, women received 43.5% of the doctorates in the most recent numbers available (Statistics Canada, 2001). Data from the United States are remarkably similar: In recent years women earned 46% of the doctorates in biological sciences but only 18% of doctorates in physics (National Science Foundation, n.d.).

It is difficult to credit the general validity of women choosing science less often, of explanations such as a "chilly climate," of the "old boy network," or of the harsh requirements of a lab-intensive environment when the number of women in biological sciences is so high. Although biological sciences certainly require some mathematical ability, it is not at a level required for physics. The pattern fits well both with the distribution of math talent across the sexes and with the greater interest in person-oriented activities in women, or in animate versus inanimate objects. Women's higher levels of nurturance apparently can be reduced by high levels of prenatal androgens (Leveroni & Berenbaum, 1998).

Government policies in North America, and to some extent also in Europe, have chosen to ignore these practical realities. Scholarships and fellowships available exclusively to women in science are now commonplace, despite the explicit conflicting dictum from government bureaus that discrimination is forbidden. In Canada, a program called University Faculty Awards used to be competitively open to both men and women in science for several years (1986–1989). These were highly desirable awards because they were tied to faculty positions and paid the bulk of salaries for several years, as well as providing research grants. The applicants were overwhelmingly male, so despite the relatively higher success rate for women over that period, only 57 (14%) went to women and 363 (86%) to men (T. Brychey, personal communication, August 20, 1999). Since then, the awards have been labeled Women's Faculty Awards and are available *only* to women. It must logically follow that hundreds of better qualified men have been passed over in the zeal to artificially raise the numbers of women in science. It appears that Harvard University president Lawrence Summers's attempt to engage his faculty in rational discussion of the situation in science will in the end result in Harvard's playing a similar game.

The argument is often made that because women were historically disadvantaged in faculty hiring, it is reasonable now to deliberately discriminate in their favor, to correct for past practices. However, the evidence does not support the claim that women have been kept from faculty positions in science or other fields over the past 2 decades, by either deliberate or systemic discrimination. In fact, Canadian data indicate that women are overhired in university faculty positions, relative to their representation in the pool of applicants (Irvine, 1996; Kimura, 2002a; Seligman, 2001). Targeting advantages like special scholarships or grants exclusively to women in disciplines that women are not drawn to in essence bribes them to enter fields they may neither excel in nor enjoy. Preferential perks of this kind also bribe employer universities to overlook merit in hiring decisions. Why not take a woman if you can get her for free, even though there may be men who are better qualified but for whom funding for salaries must be found? Apart from the flagrant injustice to men, there are likely to be long-term negative effects on the level of scholarship in universities, and on the education that students receive.

The evidence is strong that women's lower representation in some fields of science is due to innate talent and interest differences between the sexes. We are, however, speaking of average differences, and given the substantial overlap between men and women on all cognitive functions, using sex to determine quotas of admission to any program would be a mistake. There will be women who make outstanding engineers, as there are men who will make outstanding secretaries. Not discriminating on the basis of sex is a double-edged sword, however, and requires that we accept the probability of differential representation of the sexes over a wide variety of activities.

REFERENCES

Baenninger, M., & Newcombe N. (1989). The role of experience in spatial test performance: A meta-analysis. *Sex Roles, 20,* 327–344.

Benbow, C. P. (1988). Sex differences in mathematical reasoning ability in intellectually talented preadolescents: Their nature, effects, and possible causes. *Behavioral and Brain Sciences, 11,* 169–182.

Denckla, M. B. (1974). Development of motor co-ordination in young children. *Developmental Medicine & Child Neurology, 16,* 729–741.

Feingold, A. (1996). Cognitive gender differences: Where are they and why are they there? *Learning and Individual Differences, 8,* 25–32.

Felson, R. B., & Trudeau, L. (1991). Gender differences in mathematics performance. *Social Psychology Quarterly, 54*, 113–126.

Geary, D. C., & DeSoto, M. C. (2001). Sex differences in spatial abilities among adults in the United States and China. *Evolution and Cognition, 7*, 172–177.

Halpern, D. F. (1992). *Sex differences in cognitive abilities* (2nd ed.) Hillsdale, NJ: Erlbaum.

Hampson, E., Rovet, J. F., & Altmann, D. (1998). Spatial reasoning in children with congenital adrenal hyperplasia due to 21-hydroxylase deficiency. *Developmental Neuropsychology, 14*, 299–320.

Ingram, D. (1975). Motor asymmetries in young children. *Neuropsychologia, 13*, 95–102.

Irvine, A. D. (1996). Jack and Jill and employment equity. *Dialogue, 35*, 255–291.

Irwing, P., & Lynn, R. (2005). Sex differences in means and variability on the progressive matrices in university students: A meta-analysis. *British Journal of Psychology, 96*, 505–524.

Kimura, D. (1992). Sex differences in the brain. *Scientific American, 267*, 118–125.

Kimura, D. (2000). *Sex and cognition.* Cambridge, MA: MIT Press.

Kimura, D. (2002a, January 10). Preferential hiring of women. *UBC Reports*, p. 2.

Kimura, D. (2002b). Sex hormones influence human cognitive pattern. *Neuroendocrinology Letters, 23*(Suppl. 4), 67–77.

Kramer, J. H., Delis, D. C., Kaplan, E., & O'Donnell, L. (1997). Developmental sex differences in verbal learning. *Neuropsychology, 11*, 577–584.

Leveroni, C. L., & Berenbaum, S. A. (1998). Early androgen effects on interest in infants: Evidence from children with congenital adrenal hyperplasia. *Developmental Neuropsychology, 14*, 321–340.

Levine, S. C., Huttenlocher, J., Taylor, J., & Langrock, A. (1999). Early sex differences in spatial skills. *Developmental Psychology, 35*, 940–949.

Lubinski, D., & Benbow, C. P. (1992). Gender differences in abilities and preferences among the gifted: Implications for the math–science pipeline. *Current Directions in Psychological Science, 1*, 61–66.

McGuinness, D., Olson, A., & Chapman, J. (1990). Sex differences in incidental recall for words and pictures. *Learning and Individual Differences, 2*, 263–285.

National Academy of Sciences. (2005, December 9). *Convocation on maximizing the potential of women in academe: Biological, social, and organizational contributions to science and engineering success.* Washington, DC: Committee on Women in Academic Science and Engineering.

National Science Foundation. (n.d.). *Science and engineering doctoral degrees awarded to women, by field: 1996–2003.* Retrieved August 31, 2006, from http://www.nsf.gov/statistics/wmpd/pdf/tabf-2.pdf

Newcombe, N., Bandura, M. M., & Taylor, D. G. (1983). Sex differences in spatial ability and spatial activities. *Sex Roles, 9*, 377–386.

Nettles, M. T., & Millett, C. M. (2006). *Three magic letters: Getting to Ph.D.* Baltimore: Johns Hopkins University Press.

Resnick, S. M., Berenbaum, S. A., Gottesman, I. I., & Bouchard, T. J. (1986). Early hormonal influences on cognitive functioning in congenital adrenal hyperplasia. *Developmental Psychology, 22,* 191–198.

Saucier, D. M., Green, S. M., Leason, J., MacFadden, A., Bell, S., & Elias, L. J. (2002). Are sex differences in navigation caused by sexually dimorphic strategies or by differences in the ability to use the strategies? *Behavioral Neuroscience, 116,* 403–410.

Seligman, C. (2001, April). Summary of recruitment activity for all full-time faculty at the University of Western Ontario by sex and year. *SAFS Newsletter,* p. 14.

Shea, D. L., Lubinski, D., & Benbow, C. P. (2001). Importance of assessing spatial ability in intellectually talented adolescents: A 20-year longitudinal study. *Journal of Educational Psychology, 93,* 604–614.

Statistics Canada. (2001). *Survey of earned doctorates: A profile of doctoral degree recipients* (Table 2.2: Distribution by detailed field of study and gender). Ottawa, Ontario, Canada: Author.

Wainer, H., & Steinberg, L. S. (1992). Sex differences in performance on the mathematics section of the Scholastic Aptitude Test: A bidirectional validity study. *Harvard Educational Review, 62,* 323–336.

Williams, C. L., & Meck, W. H. (1993). Organizational effects of gonadal hormones induce qualitative differences in spatial navigation. In M. Haug, R. Whalen, C. Aron, & K. Olsen (Eds.), *The development of sex differences and similarities in behavior* (pp. 175–89). Amsterdam: Kluwer Academic.

3

IS MATH A GIFT?
BELIEFS THAT PUT FEMALES AT RISK

CAROL S. DWECK

Why aren't more of our brightest females pursuing careers in math and science? I was catapulted into this issue by a strange finding. In our research, my colleagues and I were looking at how students cope with confusion when they are learning brand new material. Confusion is a common occurrence in math and science, where, unlike most verbal areas, new material often involves completely new skills, concepts, or conceptual systems. So we created a new task for students to learn, and for half of the students we placed some confusing material near the beginning (Licht & Dweck, 1984).

What we found was that bright girls did not cope at all well with this confusion. In fact, the higher the girl's IQ, the worse she did. Many high-IQ girls were unable to learn the material after experiencing confusion. This did not happen to boys. For them, the higher their IQ, the better they learned. The confusion only energized them.

These findings were all the more striking because we were working with fifth-grade students. Girls were still earning higher grades than boys in just about every subject. There was no stigma attached to girls' achievement yet. And the material and problems we gave them did not involve math—so the stereotype about females and math was not in play.

Because our high-IQ girls had done wonderfully well when they did not bump up against difficulty, what we are looking at here is not a difference in ability, but a difference in how students cope with experiences that may call their ability into question—whether they feel challenged by them or demoralized by them. Barbara Licht went on to corroborate these findings. In her subsequent research, with different tasks and different measures, she also found that bright girls—who were at the top of the heap when things went well—were vulnerable to a loss of confidence and a loss in effectiveness when they ran into difficulty (Licht, Linden, Brown, & Sexton, 1984; Licht & Shapiro, 1982). And, it is here, at the top of the ability distribution, that the gender difference in math emerges. Thus, it is possible that at least part of the emerging difference in math is a gender difference in coping with setbacks and confusion rather than a gender difference in math ability.

GOOD NEWS OR BAD NEWS?

Is this good news or bad news for the issue of females in math and science? Well, it is good news if the ability is there. Yet it could be bad news, too. After all, if bright females do not cope as well with challenges, wouldn't this mean that they are not as suited for careers in math and science, careers that involve tackling the most challenging problems known and pursuing them doggedly? This is precisely why we have worked so hard to discover what lies beneath females' greater sensitivity to setbacks. We believed that if we could discover the basis for it, we could change it. Well, we have and we can.

In our recent work, we have pinpointed a psychological basis for the vulnerability. We have also shown that interventions that address this factor shrink the gender difference in math performance—both on our tasks and in the real world.

IS MATH A GIFT OR AN EARNED ABILITY?

This work starts with students' beliefs about intellectual ability in general and math ability in particular. Do they view it as a gift—an ability that you simply have or don't have? Or do they view it as something that can be developed—something that builds on an initial ability and expands it through practice and dedication?

We had found in our past research that viewing intellectual ability as a gift (a fixed entity) led students to question that ability and lose motivation when they encountered setbacks. In contrast, viewing intellectual ability as a quality that could be developed led them to seek active and effective remedies in the face of difficulty (see Dweck, 1999).

We had also shown that these beliefs about intellectual ability predicted how well students performed across the transition to junior high—a very challenging time, when grades tend to plummet and many students turn off to school (Blackwell, Trzesniewski, & Dweck, in press). Here, we found that students who viewed their intellectual ability as something they could develop maintained their interest in learning and earned significantly higher grades than their peers who viewed intelligence as a gift—even though the two groups entered junior high with the same past grades and achievement test scores. What's more, the difference in grades increased continuously over the next 2 years.

When we look within these findings at the gender story, we see that by the end of eighth grade, there is a considerable gap between girls and boys in their math grades—but only for those students who believed that intellectual skills are a gift. When we look at students who believed that intellectual ability could be expanded, the gap is almost gone. Actually, the boys are doing a little better than their fixed ability counterparts, but the girls are doing a great deal better than their counterparts (even though, again, they entered with equal math achievement). This suggests that girls who believe that intellectual abilities are just gifts do not fare well in math, but those who think they are qualities that can be developed often do just fine.

In a similar type of study, we followed students across the first semester of their premed chemistry course at Columbia University (Grant & Dweck, 2003). This highly challenging course plays a large role in who goes on to scientific careers. Here, we found the typical male–female difference in science performance—but, again, only for students who thought of intellectual ability as a gift. For the students who thought of their intellectual skills as something they could develop, the gender difference was reversed. The female students earned the higher final grades. (As always, we equated for entering ability, in this case by controlling for their SAT scores.)

A picture was beginning to emerge, then, that not all bright females are equally vulnerable. The vulnerability seems to reside more in the ones who see their ability as something that is fixed and that can be judged from their performance, so that when they hit challenges, their ability comes into question: If you have to struggle, then you must not have the gift. If your initial grades are poor, you must not have the gift.

THE GIFT AND THE STEREOTYPE

Well, we began to think, females who believe in gifts might not only be more susceptible to setbacks, they might also be more susceptible to stereotypes. After all, stereotypes are stories about gifts—about who has them and who doesn't. So if you believe in a math gift and your environment tells you that your group does not have it, then that can be disheartening. However, if, instead, you believe that math ability can be cultivated through your efforts, then the stereotype is less credible. It also seems more like something that can be overcome: "Maybe my group hasn't had the background, experience, and encouragement in the past, but with the right effort, strategies, and teaching, we should be able to make headway."

We decided to look at this issue directly. To do this, we followed female students at Columbia University through their calculus course (Good, Dweck, & Rattan, 2005a). This course is a must for virtually all math and science careers. At the beginning of the semester, we found out whether students saw math ability as a gift or whether they saw it as something that could be developed through learning. As the semester wore on, we asked them about whether they experienced gender stereotyping in their math class, and we asked them several different times about their sense of belonging in math: When they were in a math setting, did they feel accepted, respected, and comfortable—or not?

We found that many students thought that stereotyping was alive and well in their calculus section. However, happily, this had little impact on women who viewed their math ability as something they could augment. In contrast, feeling surrounded by a negative stereotype had a strong impact on women who thought of their math ability as a gift. Over the course of the semester, their sense of belonging eroded and remained low. They no longer felt accepted and comfortable in their math environment, and as a result, we found, many did not intend to pursue math in the future.

It looks, then, as though the view of math as a gift not only can make women vulnerable to declining performance, but it can also make them susceptible to stereotypes, so that when they enter an environment that denigrates their gift, they may lose the desire to carry on in that field. In this way, we were seeing highly able women drop before our eyes—women at an elite university who began the semester with high interest in math and who could well have had major careers in math or science. That is still bad news, but at least we felt we were getting a handle on the psychology behind the vulnerability, which is the first step toward better news.

WHAT CAN BE DONE?

If a big part of the problem is that women seem to lose their confidence in the face of obstacles, how can we give them more lasting confidence? One perhaps obvious solution might be to look for opportunities to praise a female's ability—for example, to watch for occasions on which a woman has done fine work and let her know that she has high ability. This may seem obvious, but it is wrong.

In a series of studies, we have shown that praising students' ability (even after a job truly well done) has a host of undesirable consequences (Mueller

& Dweck, 1998). First, it conveys to them that their ability is a gift and makes them reluctant to take on challenging tasks that hold a risk of mistakes. Next, when these same students hit a period of difficulty, the ones who had been praised for their ability tended to lose their confidence. If their success meant they had the gift, their struggles now told them they didn't. As a result, they lose interest in pursuing the task (just like females and math) and show a sharp decline in their performance.

So, it is clear that covering females with praise for their level of ability is not the answer. Rather than instill lasting confidence, it does just the opposite. So what would work? The answer, we found, is to get at the root of the vulnerable confidence by addressing students' beliefs about the *nature* of ability.

To do this, we designed an eight-session intervention for junior high school students that taught them the idea that intellectual skills can be developed (Blackwell et al., in press, Study 2). We chose the transition to junior high for the intervention because this is a time of challenge for many students, a time of declining grades, and a time when the gender difference in math often emerges. In our intervention (related to one by Joshua Aronson; see Aronson, Fried, & Good, 2002), we taught students about the brain, how it forms new connections every time they learn, and how over time this can lead to increased intellectual skills. We also taught them how to apply this lesson to their schoolwork. Students in the control group received an eight-session intervention as well, replete with high-quality instruction in useful skills, but they did not learn about the expandable nature of intellectual skills.

Before the intervention, both groups showed sharply declining grades in math, but after the intervention the group that received the "growing ability" message showed a rapid recovery and earned significantly higher math grades than the control group. Teachers (who were unaware that there were two different interventions) singled out many more students in our experimental group to say that they showed marked changes in their motivation to learn.

What is most striking for our purpose, however, is what happened to the gender difference in math. In the control group, we observed the typical gender difference, with the girls doing substantially worse than the boys. In the experimental group, that difference almost disappeared. Both groups did well.

A similar study with similar findings was conducted by Good, Aronson, and Inzlicht (2003). They, too, conducted a "growing ability" intervention with junior high school students and found higher subsequent achievement for the experimental group than the control group (in their case in achievement test scores) and a greatly reduced gap between male and female students in math achievement test scores.

In both of these interventions, then, learning that intellectual skills could be acquired—rather than simply bestowed as a gift—led to important gains in female students' math achievement. In essence, the intervention implied: If you want these skills, you can work hard and try to develop them. Girls heard this message and appeared to heed it.

What messages do we send in our math classes, and do these subtle messages make a difference? More important, can we use these messages to help female students? We (Good, Dweck, & Rattan, 2005b) tried to examine this by teaching adolescents the same math lesson in two different ways. The lesson was a geometry lesson that contained historical information about the math geniuses (Euclid & Reimann) who originated the concepts students were learning. For some students, the "innate ability" and "natural talent" of these mathematicians were highlighted. Although this may seem like innocent enough information—just a way to make a math lesson more interesting—we wondered whether this would convey to students that math was a gift bestowed on an elect few and whether this view would make female students more vulnerable when they later encountered difficulty.

For the other students, the geometry lesson contained information about the same figures but portrayed them as people who were deeply interested in and committed to math, and who worked hard and thought deeply to arrive at their contributions. This was meant to convey the idea that, whereas some people may reach the heights of proficiency (even genius), math ability is something that reaches fruition through effort.

After their lesson, students were confronted with a challenge: They were given a difficult math test, one that was said to measure their mathematical ability. When female students had received the lesson that portrayed math as a gift and then experienced this difficulty, they did significantly more poorly than their male counterparts. It was a gift and they, females, must not have it.

However, when female students got the lesson that conveyed the idea that math skills are developed, they equaled their male counterparts. The difficulty did not undermine their confidence or performance. Thus, it is clear that the messages we send in educational settings really matter, and that through our messages we can help female students perform up to their potential.

WHAT DOES THIS MEAN?

We have seen that viewing intellectual or mathematical abilities as a gift can create vulnerability in females. It makes them susceptible to a lowered sense of belonging, to a loss of confidence, and to decrements in performance in the face of difficulty and in the presence of stereotypes. However, most important, we have also seen that sending a message that these abilities can be developed can alleviate the vulnerability. Female students who heard this message—whether through an intervention or through a lesson that portrayed mathematical ability in this way—remained on a par with male students in terms of their math grades, their achievement test scores, and their performance on a very challenging math exam.

As a society, what do we believe? Do we believe in natural talent that needs little work to realize itself—that outstanding ability simply expresses itself automatically? I think many people do believe this, but researchers who study great creative contributions do not. Instead, they emphasize the idea that there is no genius, no great contribution, without great effort—not for Edison, Darwin, Mozart, or virtually anyone you can name. Despite the "legends," most geniuses put in years of intensive, even obsessive, labor before their potential reached fruition and they made the contributions we know them for (Ericsson, Krampe, & Clemens, 1993; Hayes, 1989; Weisberg, 1999). In many cases, the geniuses-to-be did not even stand out from their peers when they were younger (Bloom, 1985; Israel, 1998).

Perhaps people want to believe in innate gifts over learned abilities. That way they can put high achievers on a pedestal and see them as different from others. Well, they are different from others, but I am inclined to put more value on the process that got them there than on some ability they came with. To me, it is far more admirable to achieve something than to have it handed to you. The Polgar family produced three of the most successful female chess players ever. It was not that they showed exceptional aptitude at an early age. Rather, their father decided to work with them. Says one of the sisters, "My father believes that innate talent is nothing, that [success] is 99% hard work. I agree with him" (Flora, 2005, p. 82). The youngest daughter is now considered the best female chess player of all time, but she was not the one they considered the most talented: "Judit was a slow starter, but very hardworking" (p. 82).

One of the most damaging aspects of the "gift" mentality is that it makes us think we can know in advance who has the gift. This, I believe, is what makes us try to identify groups who have it and groups who don't—as in "boys have it and girls don't." Can anyone say for sure that there is not some gift that makes males better at math and science? What we *can* say is that many females have all the ability they need for successful careers in math-related and scientific fields and that the idea of the "gift that girls don't have" is likely to be a key part of what is keeping them from pursuing those careers.

Some say it is important to have an open, public dialogue about inherent differences in abilities and that this should not be a topic that is off-limits to scientific inquiry. Who can disagree with the assertion that it is good to have an open dialogue? Yet, we have seen that some views can harm people by telling them—in advance—that they don't have the skills and that they don't belong. I believe that the public dialogue and the scientific inquiry are best directed not at deciding who has math and science ability and who does not, but rather at how best to foster those abilities.

REFERENCES

Aronson, J., Fried, C., & Good, C. (2002). Reducing the effects of stereotype threat on African American college students by shaping theories of intelligence. *Journal of Experimental Social Psychology, 38*, 113–125.

Blackwell, L. S., Trzesniewski, K., & Dweck, C. S. (in press). Implicit theories of intelligence predict achievement across an adolescent transition: A longitudinal study and an intervention. *Child Development.*

Bloom, B. S. (1985). *Developing talent in young people.* New York: Ballantine.

Dweck, C. S. (1999). *Self-theories: Their role in motivation, personality, and development.* Philadelphia: Psychology Press.

Ericsson, K. A., Krampe, R., & Clemens, T. (1993). The role of deliberate practice in expert performance. *Psychological Review, 103*, 363–406.

Flora, C. (2005, July/August). The grandmaster experiment. *Psychology Today,* 74–84.

Good, C., Aronson, J., & Inzlicht, M. (2003). Improving adolescents' standardized test performance: An intervention to reduce the effects of stereotype threat. *Journal of Applied Developmental Psychology, 24*, 645–662.

Good, C., Dweck, C. S., & Rattan, A. (2005a). [An incremental theory decreases vulnerability to stereotypes about math ability in college females]. Unpublished data, Columbia University, New York.

Good, C., Dweck, C. S., & Rattan, A. (2005b). [Portraying genius: How fixed vs. malleable portrayal of math ability affects females' motivation and performance]. Unpublished data, Columbia University, New York.

Grant, H., & Dweck, C. S. (2003). Clarifying achievement goals and their impact. *Journal of Personality and Social Psychology, 85*, 541–553.

Hayes, J. R. (1989). Cognitive processes in creativity. In J. A. Glover, R. R. Ronning, & C. R. Reynolds (Eds.), *Handbook of creativity* (pp. 135–145). New York: Plenum Press.

Israel, P. (1998). *Edison: A life of invention.* New York: Wiley.

Licht, B. G., & Dweck, C. S. (1984). Determinants of academic achievement: The interaction of children's achievement orientations with skill area. *Developmental Psychology, 20*, 628–636.

Licht, B. G., Linden, T., Brown, D., & Sexton, M. (1984, August). *Sex differences in achievement orientation: An "A" student phenomenon?* Paper presented at the 92nd Annual Convention of the American Psychological Association, Toronto, Ontario, Canada.

Licht, B. G., & Shapiro, S. H. (1982, August). *Sex differences in attributions among high achievers.* Paper presented at the 90th Annual Convention of the American Psychological Association, Washington, DC.

Mueller, C. M., & Dweck, C. S. (1998). Intelligence praise can undermine motivation and performance. *Journal of Personality and Social Psychology, 75,* 33–52.

Weisberg, R. W. (1999). Creativity and knowledge: A challenge to theories. In R. J. Sternberg (Ed.), *Handbook of creativity* (pp. 226–250). Cambridge, England: Cambridge University Press.

4

SEX, MATH, AND SCIENCE

ELIZABETH S. SPELKE AND ARIEL D. GRACE

Harvard University's president, Lawrence Summers, initiated a public discussion of three factors that might account for the underrepresentation of women in mathematics, science, and engineering (Summers, 2005). First, sex differences in motivation may produce more men who are drawn to the single-minded pursuit of knowledge. Second, sex differences in cognition may yield more men who are capable of mathematical and scientific thinking at the highest levels. Third, discrimination may cause men to have more favorable career outcomes in these fields. In this chapter, we review research that bears on each factor.

We begin with cognitive sex differences and consider three popular claims for greater male aptitude for math and science. First, boys may be inherently more focused on objects, and girls on people. Second, boys and men may be intrinsically more gifted at numerical and spatial reasoning. Third, males may show greater variability in cognitive capacities, yielding more men than women with the best scientific minds. We believe research casts doubt on all these claims.

ARE MALES MORE ORIENTED TO OBJECTS?

The first claim is gaining new currency from the work of Simon Baron-Cohen (2003). According to Baron-Cohen, males are innately predisposed to learn about objects and mechanical relationships, and this predisposition leads more boys to become *systemizers*. Females, in contrast, are innately predisposed to learn about people and their emotions, and this predisposition leads more girls to become *empathizers*. Because systemizing is at the heart of math and science, more boys develop the cognitive skills that those fields require.

The tidal wave of interest in Baron-Cohen's thesis will surprise those who have followed the literature on the development of sex differences. The classic text in this field, by Maccoby and Jacklin (1974), found many differences between girls and boys but no difference in their focus on people versus objects. More recent studies using the statistical technique of meta-analysis, which allows investigators to aggregate information systematically across experiments, provide further evidence that boys and girls are equally interested in objects and people (Hyde, 2005). Moreover, a wealth of research has investigated the early development of knowledge about people and objects. Consistent with the conclusions of Maccoby and Jacklin, this research provides evidence that male and female infants and toddlers perceive objects and learn about mechanical relationships in strikingly convergent ways (Spelke, 2005). The few reported sex differences tend to go against the thesis that males are more focused on object mechanics. In infancy, for example, females learn a bit earlier than males that the distance an object moves depends on the force with which it is hit (Baillargeon, Kotovsky, & Needham, 1995). Common paths of learning continue through the preschool years. For example, male and female toddlers learn at indistinguishable rates how to fit blocks into holes and build towers (Keen, 2005; Ornkloo & von Hofsten, in press). This research provides no evidence that boys engage with objects more intensely or systematically. A similar body of work, beyond our scope, suggests that girls have no edge in learning about people and their mental states.

We believe this research supports an important conclusion for psychologists, educators, and parents. Young boys and girls do not divide up the labor of understanding the world. Instead, infants and toddlers of both sexes engage both with objects and with people, and they learn about objects and people with equal success. Similar conclusions hold throughout development (Hyde, 2005).

ARE MALES BETTER MATHEMATICIANS?

To evaluate whether males have greater inherent talent for mathematics, we first must consider how best to define mathematical talent. Formal mathematics is a recent accomplishment in evolutionary time: Nonhuman animals do not count or engage in symbolic calculation, and neither do modern humans in some remote societies (Gordon, 2004; Pica, Lemer, Izard, &

Dehaene, 2004). If there is a biological basis for human mathematical reasoning, it must depend on systems that evolved for other purposes.

Research from the intersecting fields of cognitive neuroscience, neuropsychology, cognitive psychology, and cognitive development provides evidence for five "core systems" at the foundations of mathematical reasoning (Dehaene, Izard, Pica, & Spelke, 2006; Feigenson, Dehaene, & Spelke, 2004; Newcombe & Huttenlocher, 2000). One system serves to represent small exact numbers of objects up to three. A second system represents large, approximate numerical magnitudes, such as about 20. Both these systems emerge in human infancy and remain functional throughout life. A third system of natural number concepts, such as seven and 29, emerges when children learn verbal counting. The last two systems represent the geometry of the surrounding surface layout and its landmarks; they are first seen when toddlers begin to navigate.

Research on the development of these five systems provides no evidence for sex differences (Spelke, 2005). Male and female infants show largely equal abilities to represent small, exact numbers of objects and events; the single reported sex difference favored females (vanMarle, 2004). No sex differences have been found in infants' or young children's representations of large approximate numerical magnitudes. Studies with large samples have found high variability in the pace of children's development of natural number concepts, but no hint of male superiority.

The systems of geometry-based and landmark-based navigation have been investigated by allowing preschool children to search for objects in a room of a given shape (e.g., a rectangle) with one or more landmarks (e.g., a colored wall), after they are disoriented by slow turning. To find the objects, children must rotate themselves back to where an object was hidden. Although these tasks are similar to tests of mental rotation, which show sex differences in adults, 2- and 4-year-old boys and girls perform them equally well (Hermer & Spelke, 1994). These and other findings suggest there are no sex differences in the primary abilities underlying mathematics (Geary, 1996).

Sex differences do emerge at older ages, on complex tasks that can be approached in multiple ways. For example, navigation through complex environments can be guided either by the geometry of the surrounding layout or by landmarks. Females tend to rely on landmarks and males on geometry. Moreover, the shapes of objects can be compared either by rotating one object into registration with another or by singling out individual features that distinguish them. More males do the former and more females do the latter. Mathematical word problems on school assessment tests often are amenable to multiple solution strategies. Girls are more apt to solve these problems by algebraic rules and boys by spatial reasoning. Because one strategy may yield faster solutions than another for any given type of problem, males and females show differing profiles on some standardized tests. The differences are not captured by the generalization that women are "verbal" and men are "mathematical" or "spatial," because each sex shows an advantage on some tests of verbal, mathematical, and spatial ability. Nevertheless, it is possible that one of these profiles is better suited to learning or performing high-level mathematics.

To evaluate this possibility, one faces a central question: By what yardstick can we compare men's and women's aptitude for math? Some people appeal to performance on the quantitative portion of the Scholastic Assessment Test (the SAT–M) for an answer. Males score higher on current versions of this test, leading many psychologists and public commentators to conclude that males have higher aptitude for mathematics (Cronin, 2005; Pinker, 2002). Appeal to the SAT–M, however, raises a problem of circularity. The SAT–M is composed of many different types of items, some of which are solved faster by the strategies females tend to favor and others by the strategies males tend to favor. By suitable choice of items, one could create a test favoring either gender. Testmakers must decide, therefore, what mix of items fairly represents the mathematical abilities of men and women. Books are devoted to this question (e.g., Gallagher & Kaufman, 2005), with consensus on one point: The only way to create a gender-fair test is to develop an independent understanding of the nature of mathematical aptitude and its distribution across the sexes. Performance on the SAT–M cannot, in itself, provide that understanding.

We suggest two ways to gain this understanding. First, mathematical aptitude can be analyzed into its component systems, and the functioning of each system can be studied in male and female infants, children, and adults. To date, such comparisons suggest no sex differences in mathematical aptitude. Second, mathematical aptitude can be studied in adults, by presenting male and female students of equal educational backgrounds with new mathematical material, over the extended time scales on which mathematicians work. If males and females differ in their aptitude for mathematics, then one sex should master this material more effectively.

This experiment takes place, on a grand scale, in high school and college mathematics classes. In U.S. high schools, girls and boys now take equally many math classes, including the most advanced ones, and girls get better grades. In U.S. colleges, women earn almost half of all bachelor's degrees in mathematics, and men and women get equal grades, equating for educational institutions and mathematics classes (Gallagher & Kaufman, 2005). Because males outscore females on the SAT–M, these findings indicate that the SAT–M systematically underpredicts the college mathematics performance of women, in relation to men. This systematic underprediction of female performance is widely acknowledged in the testing literature (e.g., Gallagher & Kaufman, 2005) but is seldom mentioned in public discussions of cognitive sex differences (e.g., Cronin, 2005; Pinker, 2002). The outcome of this large-scale experiment in nature gives us every reason to conclude that men and women have equal talent for mathematics.

MORE MALE GENIUSES?

The third claim for a male cognitive advantage in academic math and science asserts that males show greater variability in mathematical aptitude, yielding more men in the talented pool from which mathematicians and scientists are selected. Evidence supporting this claim comes from research by Benbow and her collaborators, focusing on mathematically precocious youths who were screened at adolescence, placed in intensive accelerated programs, and followed through high school, college, and beyond (Benbow & Stanley, 1983). When students first were screened by the SAT–M, there were many more high-scoring boys. Although this disparity has decreased in recent years, there still are more boys gaining the highest scores in large, talented samples. These findings seem to suggest that more males have high talent for mathematics. The suggestion can be questioned, however, because it is based entirely on a test that underpredicts women's success at mathematics.

Benbow and her collaborators collected a rich body of data on all the students in their program (not just those with the highest test scores), and these data tell a different story. In college, male and female students from the program were equally likely to major in science, with more female students in biology and more male students in physics and engineering. Moreover, equal numbers of male and female students majored in math, obtaining grades that were equally high (Lubinski & Benbow, 1992; Lubinski, Webb, Morelock, & Benbow, 2001). Despite males' higher scores on the SAT–M, the women and men in this highly talented sample showed equal mathematical ability by the most meaningful measure: They assimilated new, challenging material, in demanding mathematics classes at top-flight institutions, with equal success. Consistent with this finding, sociological studies have found that males' higher scores on the SAT–M and other similar tests do not account for sex differences in men's and women's progress to careers in mathematics, science, or engineering (Xie & Shauman, 2003).

In summary, research provides evidence that males and females are equally systematic in their thinking, they are equally endowed with the core cognitive abilities that support mathematical learning and reasoning, and they master new mathematical material equally well, both on average and in highly selected samples. We conclude that cognitive sex differences do not account for the preponderance of males on mathematics and science faculties. Let us briefly consider whether discrimination, or biological differences in motives, could explain this disparity.

GENDER BIAS AND DISCRIMINATION?

The literature on gender discrimination is large, varied, and deserving of broader treatment than we can provide here (Banaji, 1993; Spencer, Steele, & Quinn, 1999). We consider just one question: Do gender stereotypes influence the ways in which males and females are perceived as potential or practicing scientists, and do biased perceptions lead to discrimination against women in science?

We begin with research on parents' perceptions of their children. Soon after most new parents' first question—is it a boy or a girl?—has been answered, they have been asked to describe their newborn child. Parents in one study described baby boys as stronger, heartier, and bigger than baby girls, even though hospital records indicated that the boys and girls did not differ on these dimensions (Rubin, Provenzano, & Luria, 1974). Parents of older infants have been asked to predict how well their child would perform on a set of locomotor tasks such as crawling down a sloping ramp. Parents of sons were more confident that their child would make it down the ramp than parents of daughters, even though the male and female children then performed equally well (Mondschein, Adolph, & Tamis-LeMonda, 2000). Parents of children in elementary and middle school have reported that more sons than daughters showed high talent for mathematics and science, even though a panoply of measures revealed no objective differences between the girls and boys (Eccles, Jacobs, & Harold, 1990; Tenenbaum & Leaper, 2003). These findings suggest that parents overestimate the competence of their male children compared with their female children.

Further evidence that gender stereotypes influence perceptions comes from studies in which different groups of observers are presented with the same infants or children with varying gender labels. In some studies, parents or students view films of unfamiliar babies who are given either a male or a female name, and they are asked to describe the baby's actions, feelings, or attributes. When babies' actions are unambiguous, reports are not affected by the gender label. When babies' actions and feelings are ambiguous, in contrast, gender-labeling effects appear. In one study, observers viewed a child who was startled by a jack-in-the-box. Observers given a female label reported that the child was afraid, whereas those given a male label reported that the child was angry (Condry & Condry, 1976). In other studies, children with male names were more often described as strong, intelligent, and active, whereas those with female names were more often described as little and soft (Burnham & Harris, 1992; for review, see Stern & Karraker, 1989).

These differing perceptions suggest that adults may exhibit gender bias despite their best intentions. Parents may be committed to equal treatment of their sons and daughters, but no parent would respond in the same way to a fearful child and an angry child. Few parents, moreover, would behave equivalently regardless of whether they expect their child's ongoing actions to succeed or fail. If knowledge of gender affects adults' perceptions of children's emotions and abilities, then male and female children will elicit different reactions from their parents. Children will be treated differently, even by parents who believe, and report, that they treat their sons and daughters alike.

A growing literature in psychology and economics provides evidence that a similar bias occurs when professional men and women evaluate one another. Some studies have systematically compared the success of male and female scientists in competitions for fellowships and grants. Equating for a wide variety of objective measures of quality and productivity, women were found to be less likely than men to receive postdoctoral fellowships (Wenneras & Wold, 1997), and in some cases, they were awarded smaller research grants (Hosek et al., 2005). Other studies have investigated the effects of gender labeling on scientists' evaluations of other scientists. We give as an example a study of professors of psychology, who were asked to evaluate the written credentials of applicants for a faculty position (Steinpreis, Anders, & Ritzke, 1999). Each professor evaluated one dossier; for half the respondents, the name on this dossier was male, and for the other half, it was female. The respondents evaluated the candidate's accomplishments in research, teaching, and administration; they decided whether to recommend that the candidate be hired or granted tenure; and they indicated any reservations they felt about their recommendation.

One dossier depicted a candidate with an average record. For this candidate, the list of publications and courses was seen as more impressive when the name on the dossier was male than when it was female: Respondents given a male name judged that the candidate had higher research productivity and greater teaching experience. Perhaps in consequence, 70% of those presented with the male name recommended that the candidate be offered a tenure-track position, whereas only 45% of those presented with a female name did so. These effects were shown as strongly by female as by male respondents.

A second dossier depicted a candidate with an exceptionally strong record. For this candidate, there was no effect of gender labeling on the respondents' recommendations for hiring or tenure, consistent with the view that overt and explicit discrimination against women has largely disappeared. Nevertheless, many respondents expressed doubts about the candidate. For example, one respondent qualified his tenure evaluation by remarking that before he agreed to grant tenure, he would need to determine whether the publications resulted from the candidate's own work or from that of the advisor. Such questions are raised frequently, and legitimately, during real tenure evaluations. It is disconcerting to learn, however, that reservations were expressed four times more often when the name on the dossier was female.

These findings suggest pervasive differences in perceptions of the competence, accomplishments, and promise of male and female scientists. These differing perceptions may produce a pattern of discrimination in people with the best intentions. No doubt, the vast majority of mathematicians and scientists are committed to the principle that equally qualified men and women should have equal chances for advancement in their fields. Nevertheless, all faculty members favor candidates with greater research productivity, teaching experience, and scholarly independence. This research provides evidence that knowledge of a person's gender influences faculty members' assessment of these qualities.

In summary, a body of evidence suggests that males and females are perceived and evaluated differently from the moment of birth to the moment of tenure. These biased perceptions may create gender imbalances on math and science faculties in four ways. First, biased perceptions produce discrimination: When a group of equally qualified men and women are evaluated for jobs, more of the men will get those jobs if they are perceived to be more qualified. Second, bias and discrimination decrease the attractiveness of academic careers for women: More men will be drawn to academia, because they face better odds of getting promoted, funded, and honored. Third, biased perceptions of children and young adults may deter some female students from attempting a career in science or mathematics, as research by Eccles et al. (1990) and others indicates. Finally, people are more apt to consider professions that are populated by others who are like themselves. If the first three effects limit the numbers of women in science and mathematics, female students will be less likely to see these fields as possible or desirable careers.

Despite these forces, biases are malleable and cognitive development is robust. A consideration of recent history suggests that perceptions of men and women can change considerably within a single generation. Moreover, boys and girls now show equal capacities and achievements in science and mathematics from elementary school to college, despite the biased ways in which they are perceived and judged. Boys' and girls' equal performance underscores the conclusion that mathematical and scientific reasoning have a biological foundation that is common to males and females. This common foundation gives us reason to hope that the pervasive but malleable biases in perception can be overcome.

DO MEN AND WOMEN WANT DIFFERENT THINGS?

If men and women ever gain equal access to mathematics and science careers, the question of biological differences in motivation will assume center stage. Harvard's President Summers had suggested that inherent motivational differences between men and women—especially, the desire of women to balance work and family—sharply limit the number of women who are drawn to the long, intense workdays that science and mathematics require.

Could he be right, and could biological differences in motives propel more men toward careers in mathematics and science?

We have seen that male and female infants are equally predisposed to balance interests in people and objects, providing evidence against innate and early emerging sex difference in motives. We also know, however, that more adult women than men express reservations about high-powered science careers. Although women work longer hours than men do worldwide, often with poor compensation (Chen, Vanek, Lund, & Heinz, 2005), women's reservations about science may stem from a biologically based desire for shorter workdays and more family time. Alternatively, men and women may have equal desires for work and family, but they may face unequal chances of realizing both desires. Fewer women may be attracted to science today, because students of both sexes perceive, all too accurately, that women must work harder than men to achieve equal success. Moreover, both sexes may perceive, with some reason, that female scientists are less likely than their male counterparts to succeed at combining a family life with an academic career (National Science Foundation, 2004). Thus, talented men and women may want the same things: challenging, meaningful work and rich family lives. In present day industrialized societies, however, these goals may be harder for women to realize.

Deciding between these alternatives requires a grand experiment. The evidence that men and women have equal cognitive capacity, both in general and in the fields of science and mathematics, should be allowed to disseminate through society. Although decisions about hiring and promotion necessarily rest on the best judgments of other scientists, those judgments should be informed by an understanding of the forces that lead scientists to undervalue the accomplishments of women, in relation to men. Once budding scientists are perceived in relation to their actual capacities and accomplishments, irrespective of gender, we will be in a position to learn whether different inner voices pull talented men and women in different directions. At the present time, the outcome of this experiment cannot be predicted. We hope, however, that some future generation of scientists will get to conduct it.

REFERENCES

Baillargeon, R., Kotovsky, L., & Needham, A. (1995). The acquisition of physical knowledge in infancy. In D. Sperber, D. Premack, & A. J. Premack (Eds.), *Causal cognition: A multidisciplinary debate* (pp. 79–116). New York: Clarendon Press/ Oxford University Press.

Banaji, M. R. (1993). The psychology of gender: A perspective on perspectives. In A. E. Beall & R. J. Sternberg (Eds.), *The psychology of gender* (pp. 251–273). New York: Guilford Press.

Baron-Cohen, S. (2003). *The essential difference: The truth about the male and female brain*. New York: Basic Books.

Baron-Cohen, S. (2005, August 8). Op-Ed: The male condition. *The New York Times*, p. 15.

Benbow, C. P., & Stanley, J. C. (1983, December 2). Sex differences in mathematical reasoning ability: More facts. *Science, 222*, 1029–1031.

Burnham, D. K., & Harris, M. B. (1992). Effects of real gender and labeled gender on adults' perceptions of infants. *Journal of Genetic Psychology, 153*, 165–183.

Chen, M., Vanek, J., Lund, F., & Heinz, J. (2005). *Progress of the world's women 2005: Women, work and poverty*. New York: United Nations Development Fund for Women.

Condry, J., & Condry, S. (1976). Sex differences: A study of the eye of the beholder. *Child Development, 47*, 812–819.

Cronin, H. (2005, March 12). The vital statistics: Evolution, not sexism, puts us at a disadvantage in the sciences. *The Guardian*, p. 21.

Dehaene, S., Izard, V., Pica, P., & Spelke, E. S. (2006, January 20). Core knowledge of geometry in an Amazonian indigene group. *Science, 311*, 381–384.

Eccles, J. S, Jacobs, J. E., & Harold, R. D. (1990). Gender role stereotypes, expectancy effects, and parents' socialization of gender differences. *Journal of Social Issues, 46*, 183–201.

Feigenson, L., Dehaene, S., & Spelke, E. S. (2004). Core systems of number. *Trends in Cognitive Sciences, 8*, 307–314.

Gallagher, A. M., & Kaufman, J. C. (2005). *Gender differences in mathematics*. New York: Cambridge University Press.

Geary, D. C. (1996). Sexual selection and sex differences in mathematical abilities. *Behavioral and Brain Sciences, 19*, 229–284.

Gordon, P. (2004, October 15). Numerical cognition without words: Evidence from Amazonia. *Science, 306*, 496–499.

Hermer, L., & Spelke, E. S. (1994). A geometric process for spatial reorientation in young children. *Nature, 370*, 57–59.

Hosek, S. D., Cox, A. G., Ghosh-Dastidar, B., Kofner, A., Ramphal, N., Scott, J., et al. (2005). *Gender differences in major federal external grant programs* (Tech. Rep. No. TR-307-NSF). Santa Monica, CA: RAND Corporation.

Hyde, J. (2005). The gender similarities hypothesis. *American Psychologist, 60*, 581–592.

Keen, R. C. (2005, October). *Brain development, cognition and action in infants and toddlers*. Paper presented at the European Science Foundation Conference on Brain Development and Cognition in Human Infants, Acquafredda di Maratea, Italy.

Lubinski, D., & Benbow, C. P. (1992). Gender differences in abilities and preferences among the gifted: Implications for the math/science pipeline. *Current Directions in Psychological Science, 1*, 61–66.

Lubinski, D., Webb, R. M., Morelock, M. J., & Benbow, C. P. (2001). Top 1 in 10,000: A 10-year follow-up of the profoundly gifted. *Journal of Applied Psychology, 86,* 718–729.

Maccoby, E. E., & Jacklin, C. N. (1974). *Psychology of sex differences.* Stanford, CA: Stanford University Press.

Mondschein, E. R., Adolph, K. E., & Tamis-LeMonda, C. S. (2000). Gender bias in mothers' expectations about infant crawling. *Journal of Experimental Child Psychology, 77,* 306–316.

National Science Foundation, Division of Science Resource Statistics. (2004). *Gender differences in the careers of academic scientists and engineers: A literature review* (Special. Rep. No. NSF-04-323). Arlington, VA: Author.

Newcombe, N. S., & Huttenlocher, J. (2000). *Making space: The development of spatial representation and reasoning.* Cambridge, MA: MIT Press.

Ornkloo, H., & von Hofsten, C. (in press). Fitting objects into holes: On the development of spatial cognition skills. *Developmental Science.*

Pica, P., Lemer, C., Izard, V., & Dehaene, S. (2004, October 15). Exact and approximate arithmetic in an Amazonian Indigene group. *Science, 306,* 499–503.

Pinker, S. (2002). *The blank slate: The modern denial of human nature.* New York: Viking.

Rubin, J. Z., Provenzano, F. J., & Luria, Z. (1974). The eye of the beholder: Parents' views on sex of newborns. *American Journal of Orthopsychiatry, 44,* 512–519.

Spelke, E. S. (2005). Sex differences in intrinsic aptitude for mathematics and science? A critical review. *American Psychologist, 60,* 950–958.

Spencer, S. J., Steele, C. M., & Quinn, D. M. (1999). Stereotype threat and women's math performance. *Journal of Experimental Social Psychology, 35,* 4–28.

Steinpreis, R. E., Anders, K. A., & Ritzke, D. (1999). The impact of gender on the review of the curricula vitae of job applicants and tenure candidates: A national empirical study. *Sex Roles, 41,* 509–528.

Stern, M., & Karraker, K. (1989). Sex stereotyping of infants: A review of gender labeling studies. *Sex Roles, 20,* 501–522.

Summers, L. (2005, January 14). *Remarks at NBER conference on diversifying the science and engineering workforce.* Retrieved April 5, 2005, from http://www.president.harvard.edu/speeches/2005/nber.html

Tenenbaum, H. R., & Leaper, C. (2003). Parent–child conversations about science: The socialization of gender inequities? *Developmental Psychology, 39,* 34–47.

vanMarle, K. (2004). *Infants' understanding of number: The relationship between discrete and continuous quantity.* Unpublished doctoral dissertation, Yale University.

Wenneras, C., & Wold, A. (1997). Nepotism and sexism in peer-review. *Nature, 387,* 341–343.

Xie, Y., & Shauman, K. (2003). *Women in science: Career processes and outcomes.* Cambridge, MA: Harvard University Press.

5

TAKING SCIENCE SERIOUSLY: STRAIGHT THINKING ABOUT SPATIAL SEX DIFFERENCES

NORA S. NEWCOMBE

Sex differences in spatial functioning are frequently the source of cocktail party conversation and entertaining cartoons. These anecdotes and images feature such figures as men who forge intrepidly into the wild armed only with a compass and an innate sense of direction and ditzy women who hold maps upside down and depend on the kindness of strangers. These figures of fun, although exaggerated, have some basis in truth. Studies using standardized tests have found support for the belief that men have strengths in the spatial domain, sometimes quite marked strengths. For example, the average American man has an ability to perform mental rotation of a three-dimensional object that exceeds that of the average American woman by half a standard deviation or more (Linn & Petersen, 1985; Voyer, Voyer, & Bryden, 1995; for the statistically uninitiated, such differences are considered moderate to large and may well be obvious to the casual observer). There are similarly substantial sex-related differences on tests of mechanical reasoning, which involve a large spatial component (Feingold, 1995). Furthermore, and consistent with remarks made in January 2005 by Lawrence Summers, the president of Harvard University, these differences are often most marked at the upper end of the distribution. That is, men outnumber women by especially large ratios at the highest levels of these abilities (Feingold, 1995; Hedges & Nowell, 1995). Lastly, in what may seem to be the final step in a devastating argument, such differences have been found to be relevant to success in science for men and women, both because mathematical ability may rest partially on spatial ability (Casey, Nuttall, & Pezaris, 1997, 2001) and also because spatial visualization is directly relevant to achievement in many scientific and technical fields, including physical science, mathematics, computer science, and engineering (Shea, Lubinski, & Benbow, 2001).

So, given these facts, was Summers correct in speculating that women may be held back from success in science partly because they lack the cognitive prerequisites? I think the answer is yes, at a superficial level, because his descriptive facts were right—although I am not sure that the proportion of variance accounted for by cognitive variables is that large in comparison with issues such as family–work conflicts. However at a deeper level, I think the answer is no, he was incorrect, because these "facts" reflexively evoke in most people two unwarranted assumptions: that any sex-related differences are biologically caused and that they are hence immutable. (The "hence" in that last sentence is itself problematic, because many biologically caused conditions can be easily addressed by interventions, as when we color our graying hair or treat our children's ear infections with antibiotics. Yet for most people there is a linkage, albeit a mistaken one, between biology and immutability.)

In this chapter, I address each assumption in two parts. For biological causation, I first present an extremely brief overview of the current state of evidence, and then I concentrate on the evolutionary psychology framework that currently provides the primary scientific basis for taking biology very seriously, pointing out its grave inconsistencies and gaps in logic. Turning to the issue of immutability, I discuss evidence that spatial ability levels can fairly easily be increased overall. I then focus on the issue of whether sex-related differences can be eliminated and whether that matters, arguing that increasing the abilities of both women and men should be the key goal.

CURRENT EVIDENCE ON CAUSATION

When I first took a look at the literature on sex-related differences in spatial ability (Newcombe, 1982), it was common to see enthusiasm for the hypothesis that there was an X-linked recessive gene that accounted for greater male performance. The idea was that women would need two "doses" of the high-ability gene to show high ability, but that men would show the heightened ability even with only one "dose" because there would be no dominant gene on the Y chromosome to mask its effect. The excitement about this idea ultimately fizzled, however, in the face of disconfirming evidence (see review by Boles, 1980). Similarly, at that same time, there was an initial love affair with the idea that men excel in the spatial domain because they reach puberty later than women (Waber, 1976), but that too waned in the light of subsequent examination (Newcombe & Bandura, 1983; Newcombe & Dubas, 1986).

Another idea about sex differences back in that day, which is actually not completely dead, was that men are better at spatial tasks because they are "more lateralized"—more likely than women to use the right hemisphere for spatial tasks. However, the data on this subject are very messy indeed (for examples, see Johnson, McKenzie, & Hamm, 2002; Rilea, Roskos-Ewoldsen, & Boles, 2004), and overall, I think there is only dubious support for the hypothesis. The most plausible current biological hypothesis concerns the effects of sex hormones, although it is not quite clear which hormones are the most relevant, and again the data are fairly murky (see review by Collaer & Hines, 1995). Reports continue to suggest that the story on hormones and spatial ability is, at best, complex (Halari et al., 2005; Hooven, Chabris, Ellison, & Kosslyn, 2004).

Before discussing sociobiology, a small digression may be in order. It is sometimes supposed that the earlier in development an effect is observed, the more likely it is to be biologically caused. This mode of reasoning is fallacious: Late-emerging effects can be maturational, and early-emerging effects can be environmentally produced, perhaps even by in utero experiences. So, we should not be overly impressed by the fact that sex-related differences appear by the time children are in preschool, kindergarten, or early elementary school (Levine, Huttenlocher, Taylor, & Langrock, 1999; Levine, Vasilyeva, Lourenco, Newcombe, & Huttenlocher, in press). The fact is notable because it can help us to target specific evidence-based interventions, but it does not tell us that differences are biological. In fact, there is evidence on early sex differences that suggests the importance of the environment in creating them. Sex differences in two spatial tasks, one involving rotation and the other involving map reading, were not observed in young elementary school children from low-income backgrounds, although we saw the standard sex differences in middle- and high-income groups (Levine et al., in press; see also Noble, Norman, & Farah, 2005, on early socioeconomic status differences). The most natural explanation of this interaction is that boys in low-income environments lack access to experiences that enhance spatial skill (e.g., computer games, puzzles, and building sets).

DO SEX-RELATED DIFFERENCES HAVE AN EVOLUTIONARY EXPLANATION?

What does the list of failed or marginal hypotheses about biological causation of spatial sex differences leave in the way of support for the general idea that the differences are an essential aspect of being male or female? Probably the strongest reason that people believe in biological causation today is not empirical at all. Rather, their belief derives from the fact that the sex differences in spatial ability story seems to fit so neatly into an evolutionary psychology framework. So it is important to take a moment to look at that framework with a critical eye.

Steven Pinker, John Tooby, Leda Cosmides, and many other proponents of evolutionary psychology often speculate about the reasons for sex-related differences in cognition, and their speculations are so often repeated that they have attained acceptance without having been subjected to searching analysis or scientific testing. I believe that there are untested assumptions, at best, and incoherence, at worst, in the sociobiological framework regarding these differences (see also Jones, Braithwaite, & Healy, 2003). (Similar problems are evident in many other domains, such as analyses of romantic attraction; David Buller's 2005 book, *Adapting Minds*, provides a "scientific detective novel" analyzing these broader problems in sociobiology in fascinating detail.)

There are actually two evolutionary explanations of spatial sex differences. Both focus on the reproductive advantage that might accrue to men for having higher spatial ability but that would not be relevant for women. One explanation focuses on Man the Hunter. Men are generally the hunters in hunting–gathering societies, and hunting seems to require spatial skill in several of its component activities, including tracking animals, aiming at them, and fashioning the weapons with which to take aim. The protein obtained from hunting helps to ensure the survival of a man's children, and prowess in hunting may also enhance a man's access to women (who wish to have their children provisioned by a skilled hunter). In addition, aiming may come in handy in any struggles for dominance with other males over sexual access to females.

However, there are problems with this story. Although Man the Hunter may need spatial skills, so does Woman the Gatherer. Gathering may require quite long trips away from home base to find various kinds of edible vegetation in their respective ripening seasons. True, animals move while vegetation sits still, but humans are far from the swiftest creatures around, and much hunting by our ancestors may have consisted of setting traps, waiting at waterholes, and so on, rather than tracking animals over meandering paths. In terms of manufacturing artifacts to use in hunting and gathering, note that spatial skill is required to weave or make baskets or pottery, as much as for fashioning arrows and spearheads.

Another explanation for sex differences focuses on the Man Who Gets Around. In this story, men need spatial skill to navigate around the territory required to have sexual access to as large a number of fertile women as possible. There is in fact beautiful observational and experimental evidence that this kind of effect is seen in voles, a small mammal that comes in two varieties (e.g., Gaulin, FitzGerald, & Wartell, 1990). One species, the prairie vole, is pair-bonded, whereas a very similar species, the meadow vole, has a mating system in which females occupy territories and, during the mating season, males make the rounds of females, attempting to reach as many as possible in time to impregnate them. Strikingly, spatial ability (assessed by the ability to navigate mazes) is equal for males and females in the pair-bonded prairie vole, but male meadow voles beat females at spatial tasks during the mating season only, when the part of their brain that supports navigation (the hippocampus) actually enlarges to meet the reproductive challenge. Yet there is one big "sticky wicket," as the English say so bewilderingly, for extending this line of explanation to apply to human beings. Human females, unlike meadow voles, live in social groups rather than occupying widely separated home territories. The skills it takes to impregnate many females probably include such abilities as charm and stealth, more than the ability to find one's way among a cluster of huts.

There is another problem for a sociobiological explanation of sex differences in spatial ability, a question that is actually sociobiologically inspired. Why would there be a sex difference, from an evolutionary point of view, in a trait that has adaptive significance for both sexes (even if it were to be slightly greater for one sex), when there is no obvious metabolic cost for that trait? Most sex-specific traits that enhance reproductions involve attributes such as growing antlers or ornamental tails, things that are cumbersome and costly for the body to produce, that males have only because they enhance their ability to do combat with other males or attract females. There is no reason both sexes should not have a cheap-to-produce trait that is useful in a wide variety of settings.

So, why is there a biologically based sex difference in spatial ability, if there is one at all? One clue comes from data that indicate (although sometimes inconsistently, as previously noted) that spatial ability fluctuates with hormone levels, within sex. In the hormone literature, women often (but not always) show better spatial skills when at the menstruation phase of their cycle. Men often show better spatial skill when they have relatively low testosterone levels. However, although correlations with hormone levels support some kind of biological explanation, they are puzzling within the sociobiological frameworks we have sketched. There is no obvious reproductive advantage for women having enhanced spatial skill at the infertile part of their cycle, and no obvious reason why spatial skill should vary at all with testosterone level for men—or if it did, why it would not be a direct rather than inverse relationship, as is true for other behavioral traits such as aggression.

Sex differences in spatial ability, if they are linked to hormones in any consistent way at all, may exist because they are accidentally linked to hormone levels—they are a "spandrel," to use Stephen Jay Gould's term. After all, hormones account both for the appearance of acne at adolescence and for the fact that acne is typically more severe for males than females, but no one would suggest that acne confers a reproductive advantage. Acne is a clear example of a sex difference that seems likely to be accidental, and spatial ability may well be another.

SPATIAL ABILITY CAN (FAIRLY EASILY) BE VASTLY IMPROVED

Now let us take a look at the issue of malleability. Even though sex differences in spatial ability are substantial, mean levels of spatial ability do not seem to be biologically fixed. Like other intellectual abilities, and possibly more so, spatial ability has increased in the past century faster than the gene can change, a phenomenon that has been called the *Flynn effect* after its discoverer (Flynn, 1987). Some time ago, I collaborated on a meta-analysis of studies done through the late 1980s that showed very clear effects of simple practice and also of training on improvements in spatial ability, and that also showed sensible gradients of the size of the effects as a function of the time devoted to practice or training and the potential of the activities to lay the foundation for generalizability (Baenninger & Newcombe, 1989). Research subsequent to the meta-analysis has supported its broad thesis. For example, it has been shown that being in school is associated with greater spatial growth in elementary school children than being on summer vacation (Huttenlocher, Levine, & Vevea, 1998) and that various manipulations can help children learn spatial tasks (Taylor, Uttal, Fisher, & Mazepa, 2001; Uttal, Fisher, & Taylor, in press).

Most recently, I have collaborated on a new meta-analysis that again shows substantial improvements in spatial skill, for both children and adults, that emerge from academic coursework, task-specific practice, musical training (not simply musical exposure—not the so-called Mozart effect), and playing computer games (Marulis, Warren, Uttal, & Newcombe, 2005). I have also been involved in two studies that gave undergraduate students extended practice or training on mental rotation. In one study, we found that, after a semester of work, people were still improving their rotation ability with no sign of leveling off (*asymptote*), that training (playing the computer game Tetris) led to greater improvement than simple practice, that training effects lasted for months and generalized to other spatial tasks (e.g., mental paperfolding tasks), and that both practice and training effects were massive—far larger than the typical sex difference (Terlecki & Newcombe, 2005). Similar data emerged from a subsequent study, this time with evidence of symmetric transfer between mental rotation and paper folding, and using stimuli that ruled out the possibility that the effects involved simply memory for the particular stimuli (Wright, Thompson, Ganis, Kosslyn, & Newcombe, 2006).

CAN SEX DIFFERENCES BE ELIMINATED AND DOES IT MATTER?

For many people, evidence that spatial skill can improve is less important than the answer to the question of whether improvements are more marked for women than for men, so that women can catch up to men and eliminate their deficit in this intellectual ability. The current evidence is somewhat equivocal, but I read it to indicate that convergence is hard to get. We did not find sex differences in gains in the early meta-analysis, we did not find that women caught up to men even with extended practice and training (Terlecki & Newcombe, 2005), and the more recent meta-analysis also failed to find reliable sex differences in gains (although there is a hint of this effect and more data are needed). Now, failure to get convergence using current methods does not necessitate concluding it is impossible—we would hardly conclude that we can never find an AIDS vaccine just because one has not yet materialized. Whether or not men and women would converge in their abilities in more supportive or more carefully designed educational environments remains to be determined.

Let us suppose for the moment, however, that women do not catch up with men even when both sexes show enormous gains. Some investigators might argue that we would expect a disparity in elite achievement in science and technical fields to remain in this case. I am not so sure. There are multiple factors that determine success at the highest levels of science, and beyond some (high) threshold, I doubt that extra increments of the same cognitive ingredients explain much variance. Thinking creatively, explaining one's data, or inspiring a research team may be pretty important as well! Remember that relatively short and easy interventions greatly improve spatial skill—actually by the equivalent of 10 IQ points. If we want to maximize the human capital available for occupations that draw on spatial skill, such as mathematics, engineering, architecture, physical science, and computer science, we would do better to concentrate on understanding how to educate for spatial skill rather than focus solely on the explanation of sex differences (Newcombe, Mathason, & Terlecki, 2002).

REFERENCES

Baenninger, M. A., & Newcombe, N. (1989). The role of experience in spatial test performance: A meta-analysis. *Sex Roles, 20,* 327–344.

Boles, D. B. (1980). X-linkage of spatial ability: A critical review. *Child Development, 51,* 625–635.

Buller, D. J. (2005). *Adapting minds.* Cambridge, MA: MIT Press.

Casey, M. B., Nuttall, R. L., & Pezaris, E. (1997). Mediators of gender differences in mathematics college entrance test scores: A comparison of spatial skills with internalized beliefs and anxieties. *Developmental Psychology, 33,* 669–680.

Casey, M. B., Nuttall, R. L., & Pezaris, E. (2001). Spatial–mechanical reasoning skills versus mathematical self-confidence as mediators of gender differences on mathematics subtests using cross-national gender-based items. *Journal for Research in Mathematics Education, 32*, 28–57.

Collaer, M. L., & Hines, M. (1995). Human behavioral sex differences: A role for gonadal hormones during early development? *Psychological Bulletin, 118*, 55–107.

Feingold, A. (1995). The additive effects of differences in central tendency and variability are important in comparisons between groups. *American Psychologist, 50*, 5–13.

Flynn, J. R. (1987). Massive IQ gains in 14 nations: What IQ tests really measure. *Psychological Bulletin, 101*, 171–191.

Gaulin, S. J., FitzGerald, R. W., & Wartell, M. S. (1990). Sex differences in spatial ability and activity in two vole species (*Microtus ochrogaster* and *M. pennsylvanicus*). *Journal of Comparative Psychology, 104*, 88–93.

Halari, R., Hines, M., Kumari, V., Mehrota, R., Wheeler, M., Ng, V., & Sharma, T. (2005). Sex differences and individual differences in cognitive performance and their relationship to endogenous gonadal hormones and gonadotrophins. *Behavioral Neuroscience, 119*, 104–117.

Hedges, L. V., & Nowell, A. (1995, July 7). Sex differences in mental test scores, variability, and numbers of high-scoring individuals. *Science, 269*, 41–45.

Hooven, C. K., Chabris, C. F., Ellison, P. T., & Kosslyn, S. M. (2004). The relationship of male testosterone to components of mental rotation. *Neuropsychologia, 42*, 782–790.

Huttenlocher, J., Levine, S., & Vevea, J. (1998). Environmental input and cognitive growth: A study using time-period comparisons. *Child Development, 69*, 1012–1029.

Johnson, B. W., McKenzie, K. J., & Hamm, J. P. (2002). Cerebral asymmetry for mental rotation: Effects of response hand, handedness and gender. *Neuroreport, 13*, 1929–1932.

Jones, C. M., Braithwaite, V. A., & Healy, S. D. (2003). The evolution of sex differences in spatial ability. *Behavioral Neuroscience, 117*, 403–411.

Levine, S. C., Huttenlocher, J., Taylor, A., & Langrock, A. (1999). Early sex differences in spatial skill. *Developmental Psychology, 35*, 940–949.

Levine, S. C., Vasilyeva, M., Lourenco, S. F., Newcombe, N. S., & Huttenlocher, J. (in press). Socioeconomic status modifies the sex difference in spatial skill. *Psychological Science*.

Linn, M. C., & Petersen, A. C. (1985). Emergence and characterization of sex differences in spatial ability: A meta-analysis. *Child Development, 56*, 1479–1498.

Marulis, L., Warren, D., Uttal, D., & Newcombe, N. (2005, October). *A meta-analysis: The effects of training on spatial cognition in children*. Paper presented at the meeting of the Cognitive Development Society, San Diego, CA.

Newcombe, N. (1982). Sex-related differences in spatial ability: Problems and gaps in current approaches. In M. Potegal (Ed.), *Spatial abilities: Development and physiological foundations* (pp. 223–250). New York: Academic Press.

Newcombe, N., & Bandura, M. M. (1983). Effects of age at puberty on spatial ability in girls: A question of mechanism. *Developmental Psychology, 19*, 215–224.

Newcombe, N., & Dubas, J. S. (1986). Individual differences in cognitive ability: Are they related to timing of puberty? In R. M. Lerner & T. T. Foch (Eds.), *Biological–psychosocial interactions in early adolescence: A life-span perspective* (pp. 249–302). Hillsdale, NJ: Erlbaum.

Newcombe, N. S., Mathason, L., & Terlecki, M. (2002). Maximization of spatial competence: More important than finding the cause of sex differences. In A. McGillicuddy-De Lisi & R. De Lisi (Eds.), *Biology, society and behavior: The development of sex differences in cognition* (pp. 183–206). Westport, CT: Ablex.

Noble, K. G., Norman, M. F., & Farah, M. J. (2005). Neurocognitive correlates of socioeconomic status in kindergarten children. *Developmental Science, 8*, 74–87.

Rilea, S. L., Roskos-Ewoldsen, B., & Boles, D. (2004). Sex differences in spatial ability: A lateralization of function approach. *Brain and Cognition, 56*, 332–343.

Shea, D. L., Lubinski, D., & Benbow, C. P. (2001). Importance of assessing spatial ability in intellectually talented young adolescents: A 20-year longitudinal study. *Journal of Educational Psychology, 93*, 604–614.

Taylor, H. A., Uttal, D. H., Fisher, J., & Mazepa, M. (2001). Ambiguity in acquiring spatial representations from descriptions compared to depictions: The role of spatial orientation. In D. Montello (Ed.), *Spatial information theory: Foundations of geographic information science* (pp. 278–291). Berlin, Germany: Springer-Verlag.

Terlecki, M. S., & Newcombe, N. S. (2005, November). *The effects of long-term practice and training on mental rotation.* Paper presented at the meeting of the Psychonomic Society, Toronto, Ontario, Canada.

Uttal, D. H., Fisher, J. A., & Taylor, H. A. (in press). Words and maps: Developmental changes in mental models of spatial information acquired from depictions and descriptions. *Developmental Science.*

Voyer, D., Voyer, S., & Bryden, M. P. (1995). Magnitude of sex differences in spatial abilities: A meta-analysis and consideration of critical variables. *Psychological Bulletin, 117*, 250–270.

Waber, D. P. (1976, May 7). Sex differences in cognition: A function of maturation rate? *Science, 192*, 572–574.

Wright, R., Thompson, W., Ganis, G., Kosslyn, S. M., & Newcombe, N. S. (2006). *Transfer effects in training mental rotation and mental paper folding.* Manuscript in preparation.

6

SEX DIFFERENCES IN PERSONAL ATTRIBUTES FOR THE DEVELOPMENT OF SCIENTIFIC EXPERTISE

DAVID S. LUBINSKI AND CAMILLA PERSSON BENBOW

Society is becoming increasingly scientific, technological, and knowledge-based, depending on the utilization and maximization of human talent and potential (Friedman, 2005). A nation's strength, both economically and civically, is now linked to what it can call forth from the minds of its citizens. Consequently, much attention is being focused on strategies for increasing the number of science, technology, engineering, and mathematics (STEM) professionals produced in the United States and possible untapped pools of talent. For policies to be effective, they need to build on knowledge about what it takes to become excellent in STEM areas. Here, we review a series of known antecedents to achieving excellence in and commitment to math and science domains. Particular focus is on the well-documented sex differences on these attributes and the implications for male versus female repre-

Support for this chapter was provided by a Research and Training Grant from the Templeton Foundation and a National Institute of Child Health and Development Grant (P30HD15052) to the John F. Kennedy Center at Vanderbilt University. A draft of this chapter benefited from comments by Kimberly Ferriman, Gregory Park, and Jonathan Wai.

sentation in STEM disciplines. We do not focus on the educational experiences and opportunities, such as appropriate developmental placement (Benbow & Stanley, 1996; Bleske-Rechek, Lubinski, & Benbow, 2004; Colangelo, Assouline, & Gross, 2004; Cronbach, 1996; Lubinski & Benbow, 2000; Stanley, 2000) or involvement in research (Lubinski, Benbow, Shea, Eftekhari-Sanjani, & Halvorson, 2001), which are important for developing talent in STEM areas; rather, we concentrate on the personal attributes that predispose individuals to pursue and achieve highly in STEM careers (Lubinski & Benbow, 1992; Lubinski, Benbow, Webb, & Bleske-Rechek, 2006; Wai, Lubinski, & Benbow, 2005).[1]

This essay is also not about enhancing the scientific literacy of the general U.S. population. That, although critically important, is a different proposition from producing outstanding STEM professionals, the topic of this essay. Through our Study of Mathematically Precocious Youth (SMPY), we have specialized in the latter (Benbow, Lubinski, Shea, & Eftekhari-Sanjani, 2000; Lubinski & Benbow, 2000, 2001; Lubinski, Benbow, et al., 2001; Lubinski et al., 2006; Wai et al., 2005; Webb, Lubinski, & Benbow, 2002) and draw on that work for this review. Focusing on the talented, as SMPY does, is appropriate, given that most STEM professionals come from those in the top 10% in ability (Hedges & Nowell, 1995).

When examining complex outcomes, such as achieving distinction in STEM, it is important to take into account all the individual differences that factor into commitment and performance and not neglect any personal attributes that are known to be important. Doing so would lead to underdetermined (incomplete) models and violate the *total evidence rule* (taking all of the relevant personal-attribute information into account; see Carnap, 1950; Lubinski, 2000, p. 433; Lubinski & Humphreys, 1997, pp. 190–195). Thus, we try not to commit this error here in reviewing specific abilities, preferences, and commitment, which all help to explain male versus female disparities. In the case of sex differences in participation in math and science, we know (and we will show here) that although the sexes do not differ in general intelligence, they do differ in their specific ability patterns, interests, and number of hours willing to devote to their careers. Studying only one class of attributes will underestimate male versus female disparities in outcomes. Thus, these attributes will be reviewed here. Moreover, it is important to keep in mind that relatively small differences in the general population (or even no mean differences, but sex differences in variability) can eventuate in disparate male versus female ratios at elite levels (Feingold, 1995), as has been found (Hedges & Nowell, 1995; Stanley, Benbow, Brody, Dauber, & Lupkowski, 1992).

---

[1]There are many different kinds of cognitive abilities (Carroll, 1993; Snow & Lohman, 1989). In our treatment of specific abilities, we focus on those that are longitudinally stable and have been shown to be related to individual differences in the development of scientific expertise.

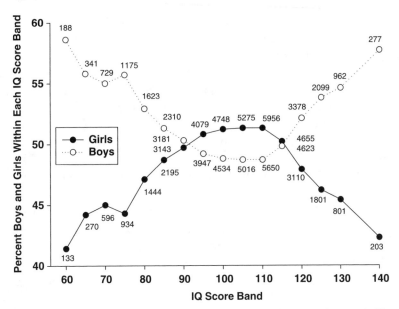

Figure 6.1. Numbers and percentages of boys and girls found within each IQ score band of the Scottish population born in 1921 and tested in the Scottish Mental Survey in 1932 at age 11. The *y*-axis represents the percentage of each sex in each 5-point band of IQ scores. Numbers beside each point represent the absolute numbers of boys and girls in each 5-point IQ score band. From "Population Sex Differences in IQ at Age 11: The Scottish Mental Survey 1932," by I. J. Deary, G. Thorpe, V. Wilson, J. M. Starr, and L. J. Whalley, 2003, *Intelligence, 31*, p. 537. Copyright 2003 by Elsevier. Reprinted with permission.

COGNITIVE ABILITIES

With regard to general ability, Deary, Thorpe, Wilson, Starr, and Whalley (2003) analyzed data collected on the complete population of 11-year-olds in Scotland in 1932 (the entire country was assessed, N = 39,343 girls and 40,033 boys). No appreciable differences in average IQ were found, but variability differences eventuated in marked male versus female ratios at the extremes of intelligence (see Figure 6.1). Thus, just as there are more boys than girls with developmental delays (e.g., mental retardation, learning disabilities), there are more highly able boys than girls. This has been observed repeatedly, with multiple samples (Arden & Plomin, 2006; Hedges & Nowell, 1995; Humphreys, 1988; Jensen, 1998; Lubinski & Dawis, 1992; Strand, Deary, & Smith, 2006).

However, specific abilities beyond general intelligence, mathematical reasoning, and spatial visualization in particular are especially critical for STEM pursuits (Humphreys, Lubinski, & Yao, 1993; Shea, Lubinski, & Benbow, 2001). Here, the sexes do differ, with males being higher in overall level as well as variability (Benbow, 1988; Hedges & Nowell, 1995; Humphreys et al., 1993; Lubinski & Humphreys, 1990; Smith, 1964).[2] Thus, there are many more males than females with high levels of these necessary (Benbow & Stanley, 1980, 1983; Gohm, Humphreys, & Yao, 1998; Lubinski & Benbow, 1992; Lubinski, Benbow, et al., 2001; Strand et al., 2006). Compounding the impact of this gender asymmetry in mathematical reasoning and spatial visualization is the tendency, even among those with more than the requisite abilities, for students to focus on their area of relative strength when choosing educational and career paths (Gottfredson, 2002, 2003; Humphreys et al., 1993; Lubinski, Benbow, et al., 2001; Lubinski, Webb, Morelock, & Benbow, 2001). Mathematically gifted individuals who are appreciably more talented in verbal than in mathematical areas are more likely to pursue careers outside of STEM. Conversely, verbally gifted individuals who are appreciably more talented in quantitative reasoning are more likely to pursue careers within STEM. That mathematically gifted females are, as a group, more verbally talented than males and more balanced in their ability profiles explains, in part, their greater attraction to intellectually demanding fields that are outside of STEM.[3]

Also taking into account the importance of spatial abilities affords an even more refined understanding of how gender disparities in STEM emerge. Shea et al. (2001), for example, tracked a group of 563 individuals representing the top 0.5% in general intellectual ability for over 20 years. They demonstrated that verbal, mathematical, and spatial abilities, all assessed in early adolescence, were related in distinctive ways to subsequent educational–vocational group membership in engineering, physical sciences, biology, humanities, law, social sciences, and business. Across developmentally sequenced

---

[2]Strand et al. (2006) published an analysis of a large and representative sample of 320,000 school pupils assessed at ages 11 through 12 in the United Kingdom. Because of the size and recency of the sample (assessed between September 2001 and August 2003), the Appendix is provided to highlight male versus female differences among the extreme scorers on measures of verbal reasoning, quantitative reasoning, and nonverbal reasoning.

[3]That specific abilities can be enhanced through learning is of course true, but a common finding is that the relationship is not linear: Those who begin with more ability typically profit more from such opportunities (Ceci & Papierno, 2005; Gagne, 2005; Jensen, 1991, p. 178; Kenny, 1975; Robinson, Abbott, Berninger, & Busse, 1996; Robinson, Abbott, Berninger, Busse, & Mukhopadhyah, 1997). For example, adolescents scoring 500 or more on SAT—Mathematics (SAT–M) or SAT—Verbal (SAT–V) before age 13 (top 1 in 200) routinely assimilate a full high school course (chemistry, English, mathematics) in 3 weeks time at summer residential programs for intellectually precocious youth; however, those scoring 700 or more (top 1 in 10,000) routinely assimilate at least twice this amount (Benbow & Stanley, 1996; Colangelo et al., 2004; Stanley, 2000). This nonlinearity is intensified by considering the full range of ability and students with developmental delays who assimilate much less than typically developing students even in the best of conditions. To the extent that all students are afforded learning opportunities individually tailored to their rate of learning, all students learn more, but individual differences in achievement are increased. Ceci and Papierno (2005, p. 149) nicely depicted this phenomenon in their subtitle: "When the 'Have Nots' Gain but the 'Haves' Gain Even More." For coming to terms with attributes of promise for exceptional achievement and creativity, it is important to keep in mind that the top 1% on essentially any ability distribution contains over one third of the ability range (e.g., for IQs, this range begins at approximately 137 and extends beyond 200); and individual differences within this 63+ IQ point range constitute differences that make a difference (Lubinski et al., 2006; Wai et al., 2005).

educational–vocational outcomes over a 20-year span, each specific ability added what statisticians term *incremental validity* (Sechrest, 1963) to the prediction of group membership relative to the other two. This is illustrated in Figure 6.2.

In Figure 6.2, longitudinal outcomes are shown for favorite and least favorite high school classes (at age 18), bachelor's degree majors (age 23), and occupations (age 33), organized around mathematical (x-axis) and verbal (y-axis) ability. For each grouping, the direction of the arrow represents whether spatial abilities (z-axis) were above (right) or below (left) the grand mean for spatial ability (A and B are within sex, C and D are combined across sex). These arrows were scaled on the same units of measurement as the SAT scores (viz., Z scores). Thus, one can envision how far apart these groups are in three-dimensional space as a function of these three abilities in standard deviation units. Across the time frames (ages 18, 23, and 33), exceptional verbal ability, relative to mathematical and spatial ability, is characteristic of group membership in the social sciences and humanities, whereas higher levels of math and spatial abilities, relative to verbal abilities, characterize group membership in engineering, math, and computer science. Engineering, for instance, is relatively high math, high spatial, and relatively low verbal. Other sciences appeared to require appreciable amounts of all three abilities. Among other things, these findings illustrate that important individual differences in ability pattern do factor into choices and outcomes, whether or not they are explicitly assessed. Indeed, spatial ability is rarely assessed. Yet, individual differences in this attribute markedly influence whether STEM domains are approached or avoided by students.

These patterns also hold for profoundly gifted participants (i.e., those scoring 700 or more on the SAT before age 13). Lubinski, Webb, et al. (2001) divided their sample of 320 profoundly gifted participants (top 1 in 10,000 students) into three groups on the basis of individual ability profiles. Two groups were "tilted" (either High-Math or High-Verbal) and one group was more intellectually uniform or "flat" (High-Flat). The High-Flat group had SAT–M and SAT–V scores that were within one standard deviation of the other. The other two groups had contrasting intellectual strengths: The High-Math group had an SAT–M score greater than one standard deviation above their SAT–V score, whereas the High-Verbal group exhibited the inverse pattern. These three ability patterns, determined from age-13 assessments, eventuated in distinct developmental trajectories. For example, age-13 assessments of specific abilities anticipated differential course preferences among these three groups in high school and college (see Figure 6.3). The High-Math group consistently preferred math and science courses relative to the humanities, whereas the inverse was true for the High-Verbal group; results among the High-Flat group were intermediate.

Lubinski, Webb, et al. (2001) also categorized the accomplishments and awards of these precocious participants into one of three clusters: Hu-

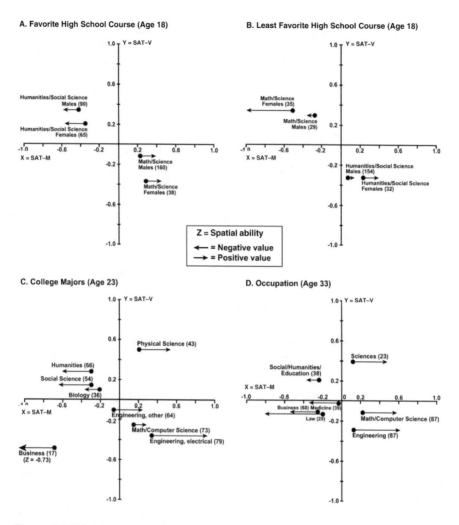

Figure 6.2. Trivariate means for (A) favorite high school course at age 18, (B) least favorite course at age 18, (C) conferred bachelor's degree at age 23, and (D) occupation at age 33. Group sample sizes are in parentheses. SAT–V = Verbal subtest of the Scholastic Assessment Test; SAT–M = Mathematical subtest of the Scholastic Assessment Test; and spatial ability = Z (a composite of two subtests of the Differential Aptitude Test: space relations + mechanical reasoning). Panels A and B are standardized within sexes, panels C and D between sexes. The large arrowhead in panel C indicates that this group's relative weakness in spatial ability is actually twice as great as that indicated by the displayed length. From "Introduction to the Special Section on Cognitive Abilities: 100 Years After Spearman's (1904) 'General Intelligence,' Objectively Determined and Measured," by D. Lubinski, 2004, *Journal of Personality and Social Psychology, 86,* p. 104. Copyright 2004 by the American Psychological Association.

manities and Arts, Science and Technology, and Other (see Figure 6.4). They then went back to ascertain whether these three clusters were occupied differentially by their three ability groups. As shown in the bottom-right panel of Figure 6.4, three-fourths of the classifiable accomplishments of High-Math

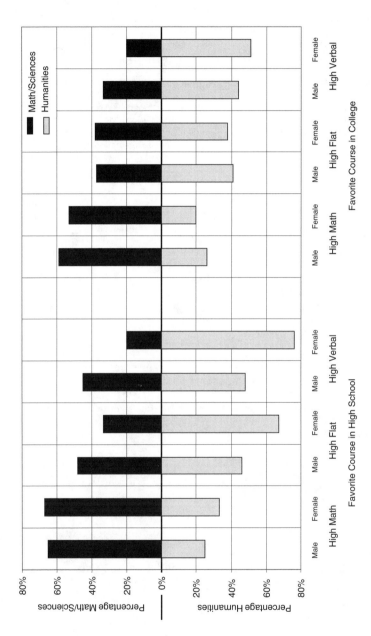

Figure 6.3. Participants' favorite course in high school and in college. Percentages in a given column do not necessarily sum to 100% because only participants indicating either math and sciences or humanities courses are displayed. Significance tests for differences among groups for favorite course are as follows: high school math and science, $\chi^2(2, N = 320) = 20.7$, $p < .0001$; college math and science, $\chi^2(2, N = 320) = 18.2$, $p < .0001$; high school humanities, $\chi^2(2, N = 320) = 36.6$, $p < .0001$; and college humanities, $\chi^2(2, N = 320) = 30.2$, $p < .0001$. From "Top 1 in 10,000: A 10-Year Follow-Up of the Profoundly Gifted," by D. Lubinski, R. M. Webb, M. J. Morelock, and C. P. Benbow, 2001, *Journal of Applied Psychology, 86,* p. 722. Copyright 2001 by the American Psychological Association.

Sciences and Technology	Humanities and Arts
Scientific publications (11) Software development (8) Inventions (4) National Science Foundation fellowship (2) Designed image correlation system for navigation for Mars Landing Program. The American Physical Society's Apker Award Graduated from MIT in 3 years at age 19 (entered at 16) with perfect (5.0) GPA and graduated from Harvard Medical School with MD at age 23. Teaching award for "Order of Magnitude Physics."	Creative writing (7) Creation of art or music (6) Fulbright award (2) Wrote proposal for a novel voting system for new South African constitution. Solo violin debut (age 13) Cincinnati Symphony Orchestra. Mellon Fellow in the humanities Presidential Scholar for creative writing Hopwood writing award Creative Anachronisms Award of Arms First place in Midreal-Medieval poetry Foreign language study fellowship International predissertation award

Other		Sciences and Technology	Humanities and Arts
Phi Beta Kappa (71) Tau Beta Pi (30) Phi Kappa Phi (14) Entrepreneurial enterprises (2) Omicron Delta Kappa Olympiad Silver Medal Finished bachelor's and master's in 4 years. Received private pilot's license in 1 month at age 17.	High Math	16	5
	High Flat	6	6
	High Verbal	7	13

Figure 6.4. Awards and special accomplishments. Numbers in brackets represent the number of participants indicating each accomplishment. All other entries represent a single individual. From "Top 1 in 10,000: A 10-Year Follow-Up of the Profoundly Gifted," by D. Lubinski, R. M. Webb, M. J. Morelock, and C. P. Benbow, 2001, *Journal of Applied Psychology, 86,* p. 725. Copyright 2001 by the American Psychological Association.

participants were in science and technology. By comparison, two-thirds of the classifiable accomplishments of High-Verbal participants were in the humanities and arts. High-Flat participants reported similar numbers of accomplishments in the sciences and humanities clusters. It is evident that ability patterns relate to the types of activities to which these individuals devoted time and effort.

These findings on course preferences, individual awards, and creative pursuits illustrate a common finding in counseling and vocational psychology, namely, that ability pattern is critical for choice (Dawis, 1992; Gottfredson, 2003). Administering one test in isolation to a group of talented adolescents is not enough to appreciate the psychological diversity among intellectually precocious youth. All three groups had exceptional SAT–M and SAT–V scores for age 13. For example, the High-Verbal group had mean SAT–M/SAT–V scores = 556/660; in contrast, the High-Flat and High-Math groups SAT–M/SAT–V means = 719/632 and 729/473, respectively.

All three groups had impressive mathematical and verbal abilities, but tilted profiles were highly related to differential development (Achter, Lubinski, & Benbow, 1996; Lubinski, Benbow, et al., 2001; Stanley et al., 1992; Strand et al., 2006). Yet there are other personal attributes highly relevant to talent development and accomplishment in STEM areas that are outside of the cognitive domain. And these too display sex differences. We turn to them next.

INTERESTS

The nearly 100-year history of research on interests is based on the truism that just because people are capable of doing something does not mean they enjoy doing it or will do it (Campbell, 1971; Dawis, 1992; Savickas & Spokane, 1999; Strong, 1943; Tyler, 1974).[4] One of the largest sex differences uncovered by psychologists studying individual differences is interest in people versus things. And, this dimension turns out also to be critical for choosing and pursuing STEM educational and career tracks.

Interests in working with people versus things can be traced back to at least Thorndike (1911), with females and males consistently displaying a mean difference of at least one standard deviation on this dimension. Girls and women, as a group, tend to prefer to learn about and work with people (or organic content), whereas boys and men, as a group, tend to prefer to learn about and work with things (or inorganic content). This dimension of individual differences routinely presents itself on educational–vocational interest inventories. Yet, current literature often fails to highlight the relevance of mean differences on this dimension for STEM pursuits, despite voluminous evidence supporting its importance (Achter et al., 1996, p. 76; Campbell, 1971; Lubinski, 2000; Lubinski, Benbow, & Ryan, 1995; Lubinski, Schmidt, & Benbow, 1996; Savickas & Spokane, 1999; Schmidt, Lubinski, & Benbow, 1998; Strong, 1943; Tyler, 1974).

How big is this sex difference today? Lippa (1998) published three studies on this robust dimension of individuality and the role it plays in personality development. Although he did not report sex differences, we were able to obtain them from him. The effect sizes (male–female differences in standard deviation units) for all three studies were greater than 1.20 (R. Lippa, personal communication, summer, 1998). This preference difference, also evident in our SMPY sample, contributes to the preponderance of females with profound mathematical gifts (viz., SAT–M $\geq$ 700, before age 13) choosing to become physicians rather than engineers and physical scientists. By contrast,

---

[4]That constellations of abilities, preferences, and conative factors are critical for coming to terms with individual differences in learning rates and occupational performance has a long history in educational, counseling, and industrial psychology (Bouchard, 1997; Corno et al., 2002; Cronbach & Snow, 1977; Dawis, 2001; Dawis & Lofquist, 1984; Snow, 1991, 1993, 1994, 1996). Scarr (1992, 1996; Scarr & McCartney, 1983) in particular has provided a developmental context for these ideas.

males with profound mathematical gifts are much more likely to become engineers and physical scientists than physicians (discussed subsequently).

A study by Webb, Lubinski, and Benbow (2002) underscores the importance of individual differences in interests for understanding educational–vocational outcomes. Webb et al. tracked 1,110 adolescents who were identified as mathematically precocious (top 1%) at age 13 and reported plans to major in math or science at the onset of their undergraduate studies. Webb et al. then compared those who eventually completed a degree in math or science with those who completed a degree in other areas. They found that more women than men eventually chose to pursue degrees in areas outside of math or science, a finding that appears negative in terms of the nation's need for more female STEM professionals. An in-depth analysis of the participants' educational, vocational, and life outcomes, however, revealed several positive findings and yielded new interpretations of the human capital that math and science domains attract.

First, Webb et al. found that individual differences in ability pattern and interests, not biological sex, surfaced as the central predictors of who actually completed a degree in math or science and who completed a degree outside of math or science. It thus appears that group status (i.e., sex) is a frail proxy variable for specific individual differences (Lubinski & Humphreys, 1997), such as ability and preference patterns, which (more centrally) guide educational–vocational choices.

Second, Webb et al. found that those who completed degrees in math or science and those who completed degrees outside of math or science showed similar levels of success, career satisfaction, and life satisfaction. For example, participants who completed their undergraduate degrees outside of math and science, regardless of sex, earned graduate degrees at comparable rates with participants within math and science; they just secured their graduate degrees in different areas. This finding mirrors other research from SMPY and other studies demonstrating that women and men with similar ability profiles achieve baccalaureate and postbaccalaureate degrees at the same rate. Yet women are more likely than men to pursue their credentials in organic fields, such as the social sciences, law, biology, and medicine. Men, in contrast, are more likely than women to pursue their credentials in inorganic fields such as engineering and the physical sciences (Achter, Lubinski, Benbow, & Eftekhari-Sanjani, 1999; Benbow et al., 2000; Lubinski, Webb, et al., 2001). This is readily seen in findings from Benbow et al.'s (2000) 20-year longitudinal follow-up of nearly 2,000 mathematically precocious youth (see Table 6.1).

CONATIVE FACTORS

It takes more than the right mix of specific abilities and interests to excel in STEM (Lubinski & Benbow, 2000, 2001; Lubinski, Benbow, et al.,

TABLE 6.1
Twenty-Year Follow-Up of Mathematically Precocious Youth

Major	Cohort 1 (Males = 840, Females = 543)						Cohort 2 (Males = 403, Females = 189)					
	Bachelor's		Master's		Doctorate		Bachelor's		Master's		Doctorate	
	M	F	M	F	M	F	M	F	M	F	M	F
Mathematics	7.5	6.3	1.0	0.9	0.4	0.2	10.3	9.7	2.2	2.1	2.2	0.5
Engineering	22.9	8.1	9.3	3.5	1.6	0.6	35.0	15.6	13.6	5.3	5.2	0.0
Computer science	7.0	4.4	3.9	2.4	1.2	0.0	10.3	2.7	6.5	0.5	2.0	0.0
Physical sciences	9.3	4.4	2.5	0.7	2.4	0.6	10.3	7.0	3.7	1.6	3.7	1.6
Biological sciences	8.1	13.5	0.5	2.4	1.1	1.3	5.8	9.6	0.5	1.6	2.2	2.1
Medicine/ health	0.7	7.7	0.5	2.0	9.9	10.7	0.3	1.6	0.5	1.1	7.4	11.6
Social sciences	17.3	19.6	2.5	2.8	1.2	0.6	9.8	19.3	2.2	6.9	0.7	2.6
Arts/ humanities	10.1	14.8	2.1	3.7	0.5	0.6	12.3	24.8	3.9	6.3	1.5	1.6
Law					7.9	6.5					6.7	11.6
Business	10.5	12.0	12.4	10.9	0.4	0.0	2.8	4.3	9.1	8.4	0.2	0.5
Education	0.5	3.1	1.1	3.1	0.1	0.2	0.0	1.6	0.0	2.6	0.0	0.0
Other fields	3.7	5.7	2.5	4.8	0.4	0.2	4.0	5.3	3.5	4.2	0.2	0.0
Math/ inorganic science	*43.7*	*21.6*	*16.2*	*7.7*	*5.5*	*1.3*	*63.5*	*34.4*	*25.3*	*9.5*	*13.2*	*2.1*
Life science/ humanities	*33.7*	*52.6*	*5.6*	*10.5*	*12.4*	*12.7*	*27.0*	*53.7*	*7.2*	*15.9*	*11.2*	*18.0*
All majors	*86.9*	*89.5*	*36.8*	*36.1*	*26.2*	*20.7*	*95.2*	*97.3*	*43.2*	*40.2*	*31.1*	*31.9*

Note. Numbers shown are percentages. The numbers do not reflect postsecondary studies under way at the time of the follow-up (Cohort 1: 2.3% of males, 4.1% of females; Cohort 2: 5.5% of males, 9.5% of females). In the summary statistics, the numbers in italics highlight a gender-differentiating trend for math and inorganic sciences and for life sciences and the humanities: Males tend to receive more degrees in the former, females the latter. F = females; M = males. From "Sex Differences in Mathematical Reasoning Ability: Their Status 20 Years Later," by C. P. Benbow, D. Lubinski, D. L. Shea, and H. Eftekhari-Sanjani, 2000, *Psychological Science, 11,* p. 475. Copyright 2000 by Blackwell Publishing. Reprinted with permission.

2001; Tyler, 1974; Williamson, 1965). Conative variables (somewhat distinct from abilities and preferences, e.g., endurance for time on task, industriousness, zeal) are highly important but underappreciated relative to abilities and interests. Their neglect has been partly caused by the difficulty associated with measuring these personal attributes. Nevertheless, regardless of the domain of exceptionality (securing tenure at a top university, making partner at a prestigious law firm, or becoming CEO of a major organization), notable accomplishments are rarely achieved by those who work 40 hours per week or less (Eysenck, 1995; Gardner, 1995; Zuckerman, 1977). World-class performers work on average 60 to 80 hours per week. They possess zeal and exhibit passion.

Consider the remarks of Dean Simonton (1994), a leading authority on the development of eminence:

> [M]aking it big is a career. People who wish to do so must organize their whole lives around a single enterprise. They must be monomaniacs, even megalomaniacs, about their pursuits. They must start early, labor continuously, and never give up the cause. Success is not for the lazy, procrastinating, or mercurial. (Simonton, 1994, p. 181)

Consider this statement by the distinguished biologist, E. O. Wilson (1998):

> I have been presumptuous enough to counsel new Ph.D.'s in biology as follows: If you choose an academic career you will need forty hours a week to perform teaching and administrative duties, another twenty hours on top of that to conduct respectable research, and still another twenty hours to accomplish really important research. This formula is not boot-camp rhetoric. (E. O. Wilson, 1998, pp. 55–56)

Figure 6.5 is based on two questions from the SMPY 20-year follow-up of nearly 2,000 intellectually precocious youth, described in Benbow et al. (2000); at age 13, their cognitive abilities were in the top 1% of their age mates. At age 33, they were asked, first, how much they would be willing to work in their "ideal job" and, second, how much they actually do work. Lubinski (2004) graphed the results, and they are displayed in Figure 6.5. Figure 6.6 is based on the same two questions administered to our top 1 in 10,000 group and the top math and science graduate students when both samples were in their mid-30s (Lubinski et al., 2006). These figures, which represent high-ability cohorts assimilated at multiple time points over 20 years, reveal an important *noncognitive* factor for exceptional achievement, willingness to work long hours, which exhibits a wide range of individual differences and an appreciable sex difference. One only needs to imagine the differences in research productivity likely to accrue over a 5- to 10-year interval between two faculty members working 45- versus 65-hour weeks (other things being equal) to understand its possible impact. The same pattern would

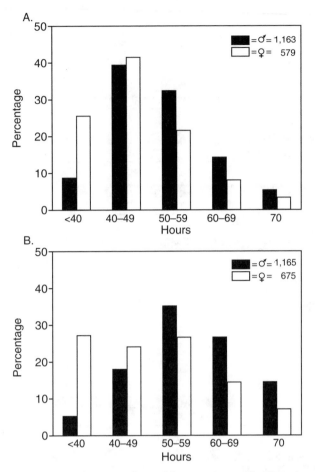

Figure 6.5. 1972–1979 talent search participants at age 33: Time devoted to work and time willing to devote to work. In the 1970s, participants were identified as having quantitative reasoning abilities in the top 1% of their age group. At age 33, they were asked (A) how many hours per week they typically work, by sex (excluding homemakers); and (B) how many hours per week they were willing to work, given their job of first choice, by sex. From "Introduction to the Special Section on Cognitive Abilities: 100 Years After Spearman's (1904) 'General Intelligence,' Objectively Determined and Measured," by D. Lubinski, 2004, *Journal of Personality and Social Psychology, 86,* p. 107. Copyright 2004 by the American Psychological Association.

emerge for advancing and achieving distinction in any other demanding pursuit (Eysenck, 1995; Gardner, 1995; Zuckerman, 1977).

These figures also reveal an interesting sex difference: An inordinate number of these exceptionally talented women were working and preferring to work 40 hours or less per week. These data fit with a number of reports in the popular press indicating that many women graduating from elite colleges are opting out of the career track, preferring to become stay-at-home moms (Story, 2005). These data also fit with normative data on hours worked (Browne, 2002).

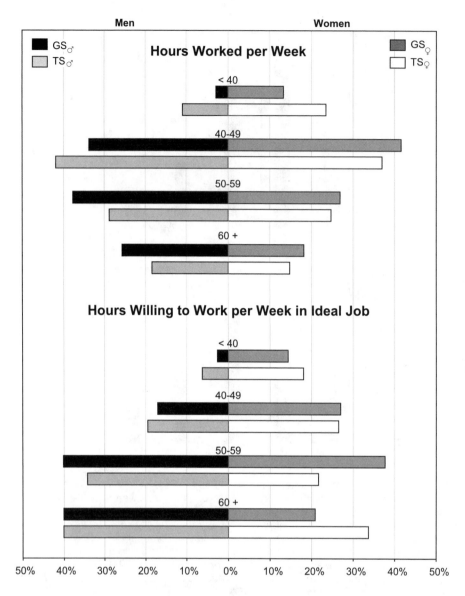

Figure 6.6. Twenty-year longitudinal follow-up, age 33, of talent search (TS) participants scoring in the top 1 in 10,000 on SAT—Mathematics or SAT—Verbal (at age 13), and a 10-year follow-up of top science, technology, engineering, and mathematics (STEM) graduate students (GS), initially identified as first- or second-year graduate students and surveyed again in their mid-30s. From "Tracking Exceptional Human Capital Over Two Decades," by D. Lubinski, C. P. Benbow, R. M. Webb, and A. Bleske-Rechek, 2006, *Psychological Science, 17,* p. 198. Copyright 2006 by Blackwell Publishing. Reprinted with permission.

It is reasonable to assume that these sex differences in time devoted to (and willing to devote to) work, if they persist, will engender large sex differences in performance and work-related outcomes with time. Indeed, Benbow et al. (2000) found that controlling for number of hours worked eliminated the commonly observed statistically significant sex differences in income.

How much time individuals are willing to devote to their careers also could engender different professional opportunities, especially in STEM areas. One aspect of STEM careers is that they are technologically rich and rapidly changing, with technical skills requiring continuous updating. More and more areas are experiencing this, but it is probably most intense in STEM. In STEM areas, taking a leave of absence for a number of years is possible, but doing so reduces significantly the probability of achieving a high-impact leadership role in subsequent employment.

CONCLUSION

Intellectually talented males and females are both achieving highly by their mid-30s. They are, however, achieving in different areas and appear to be on different developmental trajectories. Sex differences in personal attributes relevant to commitment to and excellence in STEM careers include but are not limited to ability pattern, interests, and commitment to work. These differences would predict an overrepresentation of males in STEM when males and females are free to choose how they would like to develop, other things being equal. Similarly, it is anticipated from these differences that females will be more represented in the life sciences, helping professions, and areas that place relatively greater demands on verbal skills and relatively more emphasis on a people orientation. This is exactly what SMPY and many other studies are discovering.

The findings reviewed here indicate that providing similar educational and vocational opportunities for males and females is not enough to ensure similar outcomes. When two groups differ in the ability or motivational pattern for learning and work (Corno et al., 2002; Cronbach & Snow, 1977), differential outcomes are predictable (Bleske-Rechek et al., 2004; Lubinski & Humphreys, 1997; Sackett, Schmitt, Ellingson, & Kabin, 2001). This may explain why, even though sex differences in formal math and science course-taking in high school are now negligible, women are not equally represented in engineering and the physical sciences as compared with medicine, law, biology, psychology, and many other areas (which often have a greater proportion of women). Sex differences in willingness to work long hours also have implications for how far men and women will progress, once their educational and occupational choices are made.

APPENDIX 6.1
MALE AND FEMALE DIFFERENCES AMONG EXTREME SCORERS ON COGNITIVE ABILITY TESTS

Strand et al. (2006) analyzed scores from four subtests of the Cognitive Abilities Test (CAT) from a large and representative sample of pupils in the United Kingdom (see Figure A6.1). Measures are scaled in stanines: 1 (bottom 4%), 2 (next 7%), 3 (next 12%), 4 (next 17%), 5 (middle 20%), 6 (next 17%), 7 (next 12%), 8 (next 7%), and 9 (top 4%). The sample comprised over 320,000 students, ages 11–12 years (between September 2001 and August 2003). The CAT includes separate nationally standardized tests for Verbal Reasoning (VR), Quantitative Reasoning (QR), and Nonverbal Reasoning (NVR). The mean VR score for girls was 2.2 standard score points higher than the mean for boys, but only 0.3 standard points in favor of girls for NVR, and 0.7 points in favor of boys for QR. However, for all three tests there were substantial sex differences in the standard deviation of scores, with greater variance among boys. Boys were overrepresented compared with girls at both the top and the bottom extremes for all tests, with the exception of the top 10% in verbal reasoning. On some verbal tests, more girls than boys are found at the extremes (as is found here), but results are mixed for this specific ability (cf. Lubinski & Dawis, 1992; Stanley et al., 1992).

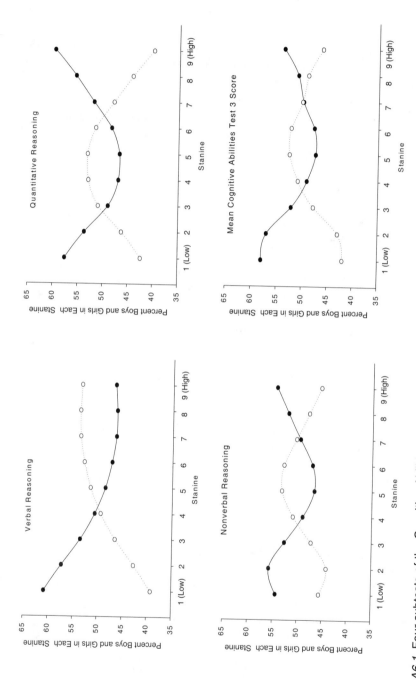

Figure A6.1. Four subtests of the Cognitive Abilities Test. From "Sex Differences in Cognitive Abilities Test Scores: A UK National Picture," by S. Strand, J. I. Deary, and P. Smith, 2006, *British Journal of Educational Psychology, 76*, p. 473. Copyright 2006 by the British Psychological Society. Reprinted with permission.

REFERENCES

Achter, J. A., Lubinski, D., & Benbow, C. P. (1996). Multipotentiality among intellectually gifted: "It was never there and already it's vanishing." *Journal of Counseling Psychology, 43*, 65–76.

Achter, J. A., Lubinski, D., Benbow, C. P., & Eftekhari-Sanjani, H. (1999). Assessing vocational preferences among gifted adolescents adds incremental validity to abilities: A discriminant analysis of educational outcomes over a 10-year interval. *Journal of Educational Psychology, 91*, 777–786.

Arden, R., & Plomin, P. (2006). Sex differences in variance of intelligence across childhood. *Personality and Individual Differences, 41*, 39–48.

Benbow, C. P. (1988). Sex differences in mathematical reasoning ability among the intellectually talented: Their characterization, consequences, and possible explanations. *Behavioral and Brain Sciences, 11*, 225–232.

Benbow, C. P., Lubinski, D., Shea, D. L., & Eftekhari-Sanjani, H. (2000). Sex differences in mathematical reasoning ability: Their status 20 years later. *Psychological Science, 11*, 474–480.

Benbow, C. P., & Stanley, J. C. (1980, December 12). Sex differences in mathematical ability: Fact or artifact? *Science, 210*, 1262–1264.

Benbow, C. P., & Stanley, J. C. (1983, December 2). Sex differences in mathematical reasoning ability: More facts. *Science, 222*, 1029–1031.

Benbow, C. P., & Stanley, J. C. (1996). Inequity in equity: How "equity" can lead to inequity for high-potential students. *Psychology, Public Policy, and Law, 2*, 249–292.

Bleske-Rechek, A., Lubinski, D., & Benbow, C. P. (2004). Meeting the educational needs of special populations: Advanced Placement's role in developing exceptional human capital. *Psychological Science, 15*, 217–224.

Bouchard, T. J., Jr. (1997). Genetic influence on mental abilities, personality, vocational interests, and work attitudes. *International Review of Industrial and Organizational Psychology, 12*, 373–395.

Browne, K. R. (2002). *Biology at work: Rethinking sexual equality*. New Brunswick, NJ: Rutgers University Press.

Campbell, D. P. (1971). *Handbook for the Strong Vocational Interest Blank*. Stanford, CA: Stanford University Press.

Carnap, R. (1950). *Logical foundations of probability*. Chicago: University of Chicago Press.

Carroll, J. B. (1993). *Human cognitive abilities: A survey of factor-analytic studies*. Cambridge, England: Cambridge University Press.

Ceci, S. J., & Papierno, P. B. (2005). The rhetoric and reality of gap closing: When the "have nots" gain but the "haves" gain even more. *American Psychologist, 60*, 149–160.

Colangelo, N., Assouline, S. G., & Gross, M. U. M. (Eds.). (2004). *A nation deceived: How schools hold back America's brightest students*. Iowa City: University of Iowa Press.

Corno, L., Cronbach, L. J., Kupermintz, H., Lohman, D. F., Mandinach, E. B., Porteus, A. W., & Talbert, J. E. (Eds.). (2002). *Remaking the concept of aptitude: Extending the legacy of Richard E. Snow*. Mahwah, NJ: Erlbaum.

Cronbach, L. J. (1996). Acceleration among the Terman males: Correlates in midlife and after. In C. P. Benbow & D. Lubinski (Eds.), *Intellectual talent: Psychometric and social issues* (pp. 179–191). Baltimore: Johns Hopkins University Press.

Cronbach, L. J., & Snow, R. E. (1977). *Aptitudes and instructional methods*. New York: Irvington Publishers.

Dawis, R. V. (1992). The individual differences tradition in counseling psychology. *Journal of Counseling Psychology, 39*, 7–19.

Dawis, R. V. (2001). Toward a psychology of values. *The Counseling Psychologist, 29*, 458–465.

Dawis, R. V., & Lofquist, L. H. (1984). *A psychological theory of work adjustment: An individual differences model and its applications*. Minneapolis: University of Minnesota Press.

Deary, I. J., Thorpe, G., Wilson, V., Starr, J. M., & Whalley, L. J. (2003). Population sex differences in IQ at age 11: The Scottish Mental Survey 1932. *Intelligence, 31*, 533–542.

Eysenck, H. J. (1995). *Genius: The natural history of creativity*. Cambridge, England: Cambridge University Press.

Feingold, A. (1995). The additive effects of differences in central tendency and variability are important in comparisons between groups. *American Psychologist, 50*, 5–13.

Friedman, T. L. (2005). *The world is flat*. New York: Farrar, Straus & Giroux.

Gagne, F. (2005). From noncompetence to exceptional talent: Exploring the range of academic achievement within and between grade levels. *Gifted Child Quarterly, 49*, 139–153.

Gardner, H. (1995). *Creating minds*. New York: Basic Books.

Gohm, C. L., Humphreys, L. G., & Yao, G. (1998). Underachievement among spatially gifted students. *American Educational Research Journal, 35*, 515–531.

Gottfredson, L. S. (2002). Assess and assist individuals, not sexes. *Issues in Education, 8*, 39–47.

Gottfredson, L. S. (2003). The challenge and promise of cognitive career assessment. *Journal of Career Assessment, 11*, 115–135.

Hedges, L. V., & Nowell, A. (1995, July 7). Sex differences in mental test scores, variability, and numbers of high-scoring individuals. *Science, 269*, 41–45.

Humphreys, L. G. (1988). Sex differences in variability may be more important than sex differences in means. *Behavioral and Brain Sciences, 11*, 195–196.

Humphreys, L. G., Lubinski, D., & Yao, G. (1993). Utility of predicting group membership and the role of spatial visualization in becoming an engineer, physical scientist, or artist. *Journal of Applied Psychology, 78*, 250–261.

Jensen, A. R. (1991). Spearman's g and the problem of educational equality. *Oxford Review of Education, 17,* 169–187.

Jensen, A. R. (1998). *The g factor: The science of mental ability (human evolution, behavior, and intelligence).* Westport, CT: Praeger Publishers.

Kenny, D. A. (1975). A quasi-experimental approach to assessing treatment effects in the nonequivalent control group design. *Psychological Bulletin, 82,* 345–362.

Lippa, R. (1998). Gender-related individual differences and the structure of vocational interests: The importance of the people–things dimension. *Journal of Personality and Social Psychology, 74,* 996–1009.

Lubinski, D. (2000). Scientific and social significance of assessing individual differences: "Sinking shafts at a few critical points." *Annual Review of Psychology, 51,* 405–444.

Lubinski, D. (2004). Introduction to the special section on cognitive abilities: 100 years after Spearman's (1904) "'General intelligence,' objectively determined and measured." *Journal of Personality and Social Psychology, 86,* 96–111.

Lubinski, D., & Benbow, C. P. (1992). Gender differences in abilities and preferences among the gifted. *Current Directions in Psychological Science, 1,* 61–66.

Lubinski, D., & Benbow, C. P. (2000). States of excellence. *American Psychologist, 55,* 137–150.

Lubinski, D., & Benbow, C. P. (2001). Choosing excellence. *American Psychologist, 56,* 76–77.

Lubinski, D., Benbow, C. P., & Ryan, J. (1995). Stability of vocational interests among the intellectually gifted from adolescence to adulthood: A 15-year longitudinal study. *Journal of Applied Psychology, 80,* 90–94.

Lubinski, D., Benbow, C. P., Shea, D. L., Eftekhari-Sanjani, H., & Halvorson, M. B. J. (2001). Men and women at promise for scientific excellence: Similarity not dissimilarity. *Psychological Science, 12,* 309–317.

Lubinski, D., Benbow, C. P., Webb, R. M., & Bleske-Rechek, A. (2006). Tracking exceptional human capital over two decades. *Psychological Science, 17,* 194–199.

Lubinski, D., & Dawis, R. V. (1992). Aptitudes, skills, and proficiencies. In M. D. Dunnette & L. M. Hough (Eds.), *Handbook of industrial/organizational psychology* (2nd ed., Vol. 3, pp. 1–59). Palo Alto, CA: Consulting Psychologists Press.

Lubinski, D., & Humphreys, L. G. (1990). Assessing spurious "moderator effects": Illustrated substantively with the hypothesized ("synergistic") relation between spatial visualization and mathematical ability. *Psychological Bulletin, 107,* 385–393.

Lubinski, D., & Humphreys, L. G. (1997). Incorporating general intelligence into epidemiology and the social sciences. *Intelligence, 24,* 159–201.

Lubinski, D., Schmidt, D. B., & Benbow, C. P. (1996). A 20-year stability analysis of the Study of Values for intellectually gifted individuals from adolescence to adulthood. *Journal of Applied Psychology, 81,* 443–451.

Lubinski, D., Webb, R. M., Morelock, M. J., & Benbow, C. P. (2001). Top 1 in 10,000: A 10-year follow-up of the profoundly gifted. *Journal of Applied Psychology, 86,* 718–729.

Robinson, N. M., Abbott, R. D., Berninger, V. W., & Busse, J. (1996). The structure of abilities in mathematically precocious young children: Gender similarities and differences. *Journal of Educational Psychology, 88,* 341–352.

Robinson, N. M., Abbott, R. D., Berninger, V. W., Busse, J., & Mukhopadhyah, S. (1997). Developmental changes in mathematically precocious young children. *Gifted Child Quarterly, 41,* 145–158.

Sackett, P. R., Schmitt, N., Ellingson, J. E., & Kabin, M. B. (2001). High-stakes testing in employment, credentialing, and higher education. *American Psychologist, 56,* 302–318.

Savickas, M. L., & Spokane, A. R. (1999). *Vocational interests: Meaning, measurement, and counseling use.* Pale Alto, CA: Davies-Black/Counseling Psychologists Press.

Scarr, S. (1992). Developmental theories for the 1990s: Development and individual differences. *Child Development, 63,* 1–19.

Scarr, S. (1996). How people make their own environments: Implications for parents and policy makers. *Psychology, Public Policy, and Law, 2,* 204–228.

Scarr, S., & McCartney, K. (1983). How people make their own environments: A theory of genotype → environment effects. *Child Development, 54,* 424–435.

Schmidt, D. B., Lubinski, D. S., & Benbow, C. P. (1998). Validity of assessing educational–vocational preference dimensions among intellectually talented 13-year-olds. *Journal of Counseling Psychology, 45,* 436–453.

Sechrest, L. (1963). Incremental validity: A recommendation. *Educational and Psychological Measurement, 23,* 153–158.

Shea, D. L., Lubinski, D. S., & Benbow, C. P. (2001). Importance of assessing spatial ability in intellectually talented young adolescents: A 20-year longitudinal study. *Journal of Educational Psychology, 93,* 604–614.

Simonton, D. K. (1994). *Greatness: Who makes history and why.* New York: Guilford Press.

Smith, I. M. (1964). *Spatial ability: Its educational and social significance.* London: University of London Press.

Snow, R. E. (1991). The concept of aptitude. In R. E. Snow & D. E. Wiley (Eds.), *Improving inquiry in the social sciences* (pp. 249–284). Hillsdale, NJ: Erlbaum.

Snow, R. E. (1993). *Construct validity of constructed-response tests.* In R. E. Bennett & W. C. Ward (Eds.), *Construction versus choice in cognitive measurement* (pp. 45–60). Hillsdale, NJ: Erlbaum.

Snow, R. E. (1994). A person–situation interaction theory of intelligence in outline. In A. Demetriou & A. Efklides (Ed.), *Intelligence, mind, and reasoning: Structure and development* (pp. 11–28). Amsterdam: Elsevier.

Snow, R. E. (1996). Aptitude development and education. *Psychology, Public Policy, and Law, 3/4,* 536–560.

Snow, R. E., & Lohman, D. F. (1989). Implications of cognitive psychology for educational measurement. In R. L. Linn (Ed.), *Educational measurement* (3rd ed., pp. 263–331). New York: Collier.

Stanley, J. C. (2000). Helping students learn only what they don't already know. *Psychology, Public Policy, and Law, 6*, 216–222.

Stanley, J. C., Benbow, C. P., Brody, L. E., Dauber, S., & Lupkowski, A. (1992). Gender differences on eighty-six nationally standardized aptitude and achievement tests. In N. Colangelo, S. G. Assouline, & D. L. Ambroson (Eds.), *Talent development: Proceedings from the 1991 Henry B. Jocelyn Wallace National Research Symposium on talent development.* (pp. 42–65). New York: Trillium Press.

Story, L. (2005, September 20). Many women at elite colleges set career path to motherhood. *New York Times*, p. A1.

Strand, S., Deary, I. J., & Smith, P. (2006). Sex differences in cognitive abilities test scores: A UK national picture. *British Journal of Educational Psychology, 76*, 463–480.

Strong, E. K., Jr. (1943). *Vocational interests for men and women*. Stanford, CA: Stanford University Press.

Thorndike, E. L. (1911). *Individuality*. New York: Houghton Miffin.

Tyler, L. E. (1974). *Individual differences*. Englewood Cliffs, NJ: Prentice-Hall.

Wai, J., Lubinski, D., & Benbow, C. P. (2005). Creativity and occupational accomplishments among intellectually precocious youth: An age 13 to age 33 longitudinal study. *Journal of Educational Psychology, 97*, 484–492.

Webb, R. M., Lubinski, D., & Benbow, C. P. (2002). Mathematically facile adolescents with math/science aspirations: New perspectives on their educational and vocational development. *Journal of Educational Psychology, 94*, 785–794.

Webb, R. M., Lubinski, D., & Benbow, C. P. (in press). Spatial ability: A neglected dimension in talent searches for intellectually precocious youth. *Journal of Educational Psychology*.

Williamson, E. G. (1965). *Vocational counseling*. New York: McGraw-Hill.

Wilson, E. O. (1998). *Consilience: The unity of knowledge*. New York: Knopf.

Zuckerman, H. (1977). *Scientific elite*. New York: Free Press.

7

DO SEX DIFFERENCES IN COGNITION CAUSE THE SHORTAGE OF WOMEN IN SCIENCE?

MELISSA HINES

I have spent much of my research career (about 30 years now) studying the influences of sex hormones, particularly androgens, such as testosterone, and estrogens, such as estradiol, on behavior. I am trained as a clinical psychologist and as a personality and developmental psychologist, as well as a neuroendocrinologist and neuroscientist, and I bring all of these perspectives to my work. Much of my research has investigated cognitive outcomes in individuals with genetic abnormalities causing sex-atypical hormone exposure prenatally or has evaluated the consequences of hormone treatment, either prenatally or in adulthood, on human cognition. On the basis of my 30 years of work, I believe the short answer to the question posed in the title of this chapter is "no." Innate sex differences in cognitive abilities do not cause the shortage of women in science. A somewhat longer answer follows. It draws heavily on my book, *Brain Gender* (Hines, 2004), and interested readers are referred there for a more detailed discussion of the points made here.

The term *sex difference* is used to describe an average difference between males and females, with some degree of overlap between the sexes.

Because of this overlap, it is important to have some idea of how large particular sex differences are. The size of sex differences can be described in standard deviation units (d, which equals the difference between mean scores for two groups—in this case males and females—divided by the pooled or average standard deviation). The size of the sex difference in height ($d = 2.0$) provides a familiar, and substantially larger, comparison value for cognitive sex differences ($d = 0.9$ or less). However, these sex differences are still of interest. In fact, in behavioral science research, sex differences of $0.8\ d$ or larger are considered large, those of about $0.5\ d$ are considered moderate, those of about $0.2\ d$ are considered small, and only those smaller than $0.2\ d$ are considered negligible (Cohen, 1988).

Saying that a characteristic shows a sex difference does not mean that it is innate or that it cannot be changed, just that it exists in society at the present time. However, I think that the question posed is related to whether innate or immutable differences limit women's ability to succeed in the scientific arena. Therefore, this chapter will discuss not only what sex differences exist in cognition but also whether they are likely to arise from innate or immutable factors.

The idea that innate differences between the sexes limit women's achievement is not new. In his review of scientific justifications of social inequities, Steven J. Gould (1981) noted that some eminent 19th-century scientists, including Paul Broca (neuroanatomist and founder of the Anthropological Society of Paris) and Gustave Le Bon (founder of social psychology), contended that women and Black men were less intelligent than White men because they had smaller brains. Similarly, German scholars at the time argued that they were superior to French scholars because the French had smaller brains. In regard specifically to women, Le Bon said,

> All psychologists who have studied the intelligence of women, as well as poets and novelists, recognize today that they represent the most inferior forms of human evolution and that they are closer to children and savages than to an adult, civilized man. They excel in fickleness, inconstancy, absence of thought and logic, and incapacity to reason. (Le Bon, 1879, pp. 60–61, quoted in Gould, 1981, pp. 104–105)

Le Bon and Broca were correct in saying that women have smaller brains than men. However, the meaning of this sex difference in brain size is debated. Some still argue, like Broca and Le Bon, that it causes men to be more intelligent than women (Lynn, 1999). However, others have suggested that the male brain is larger simply because the rest of the male body is. The sex difference in brain size is only about half the size of the sex difference in body size, suggesting brain size could be considered larger in women, if body size were controlled for. Perhaps more important, subtler aspects of brain architecture are probably more relevant to functional capacity than is overall size. For instance, in some regions of the cerebral cortex, neurons are packed more densely in females than in males. Similarly, compared with the male brain, the female brain has a higher percentage of gray matter, greater cortical volume, and increased glucose metabolism, suggesting greater functional activity. Thus, depending on the aspect of brain structure that is the focus (size, size adjusted for body size, neuron number, cortical volume, or percentage gray matter), one could argue that either males or females have more inherent intellectual capacity (Hines, 2004).

COGNITIVE SEX DIFFERENCES

There appears to be no sex difference in general intelligence, although some IQ measures show small sex differences favoring one sex or the other. Claims that men are more intelligent than women are not supported by existing data. Similarly, the idea that efforts to make intelligence tests gender neutral have removed a natural male advantage is incorrect; the most commonly used intelligence tests showed negligible and inconsistent sex differences even before gender-balancing efforts began (Loehlin, 2000).

Despite the lack of a sex difference in general intelligence, some specific cognitive abilities show sex difference. These differences are sometimes summarized as men excelling on spatial and math abilities, and women on verbal abilities. However, this is an oversimplification. Meta-analyses, which combine results of many studies, suggest that sex differences are limited to subtypes of these abilities. They also can vary with age. Additionally, even the largest cognitive sex differences ($d = 0.3$ to 0.9) are substantially smaller than the sex difference in height ($d = 2.0$).

With regard to verbal abilities, there is no sex difference in vocabulary ($d = -0.02$) or reading comprehension ($d = -0.03$), but females show a small to moderate advantage on some measures of verbal fluency ($d = 0.33$ to 0.53; Hyde & Linn, 1988). Such measures might require, for example, listing words that begin with specified letters (e.g., "F") or that fit certain categories (e.g., round things).

For spatial abilities, sex differences again vary for different tasks (Linn & Petersen, 1985; Voyer, Voyer, & Bryden, 1995). Perhaps the largest sex difference ($d = 0.9$) is seen on three-dimensional (3-D) mental rotations ability or the ability to rotate stimuli rapidly and accurately within the mind. These measures require indicating whether figures (e.g., drawings of blocks joined together to form 3-D shapes) are the same as one another but simply rotated, as opposed to not only being rotated but also mirror images. Males excel on this type of task, and the sex difference appears to be larger in adults than in children. Men also excel on measures of spatial perception ($d = 0.5$), which can require indicating, for example, which lines in a semicircular array have the same angle of orientation as sample lines, or drawing the water level in a tilted jar, or aligning a glowing rod to the horizontal in a darkened room without visual clues as to the true horizontal or vertical. Many other types of spatial tasks show negligible sex differences ($d < 0.2$). These include measures of spatial visualization, involving complex, multistep procedures for solution. Examples include finding hidden patterns in more complicated shapes, arranging blocks to resemble sample figures, and imaging how flat shapes could be folded to form 3-D objects.

For mathematical abilities (Hyde, Fennema, & Lamon, 1990), there is a negligible overall sex difference favoring females ($d = -0.05$) and a similar negligible sex difference in understanding of mathematical concepts ($d = -0.06$). There is also a small to negligible sex difference favoring girls in mathematical computations ($d = -0.21$), although this sex difference is absent in adulthood. Also, males have historically outperformed females on the mathematics subtest of the Scholastic Aptitude Test ($d = 0.38$) and the Graduate Record Exam ($d = 0.77$), tests used in the United States to select students for admission to bachelor's and doctoral study, respectively.

In the 1970s, an X-linked recessive gene was thought to cause sex differences in spatial abilities (Maccoby & Jacklin, 1974). This conclusion derived from several studies reporting a cross-sex-linked pattern of correlations between parents and children on spatial tasks. However, subsequent research, involving larger samples, failed to replicate the initial findings (Hines, 2004). In addition, basic research on sexual differentiation had identified gonadal hormones as the primary determinants of sex differences in the brain and behavior, and the search for innate causes of sex differences moved from genes to hormones. Currently, some believe that gonadal hormones are major determinants of sex differences in cognition and that they explain the predominance of men in certain scientific professions (Kimura, 1999). However, research findings in this area are inconsistent, and hormonal explanations of sex differences in spatial or other cognitive abilities may prove no more reliable than prior explanations based on brain size or X-linked genes.

Research on nonhuman mammals provides a basis for suggesting that hormones influence human cognition. In mammals, including humans, male fetuses and neonates have higher levels of testicular hormones, particularly testosterone, than do females. In addition, hundreds of experimental studies in nonhuman mammals have shown that the early (prenatal and neonatal) hormone environment influences brain development and that, consequently, variability in hormones during these early critical periods permanently alters behaviors that show sex differences (Goy & McEwen, 1980). For instance, treating genetic female rats with testosterone during early development alters neural structures, as well as sexual and other behaviors, including maze-learning behaviors, that show sex differences. Similar behavioral influences of testosterone have been documented in nonhuman primates, although, as yet, not specifically on maze learning. These permanent effects of the early hormone environment have been termed *organizational effects*, because the hormones influence the fundamental organization of the brain. In addition to these organizational effects, hormonal fluctuations in adulthood can influence behaviors that show sex differences. These adult influences are called *activational*, because the hormones are thought to influence behavior by activating existing neural circuits. Activational influences differ from organizational influences in being transient—they are present as long as the hormone is, and they disappear when it is withdrawn. In contrast, organizational effects are permanent, persisting even after the hormone is no longer present.

Research on human sexuality and on gender role behavior in children indicates that the prenatal hormone environment influences the development of some human behaviors that show sex differences. For example, girls with congenital adrenal hyperplasia (CAH), a genetic disorder that causes the adrenal gland to overproduce androgens, including testosterone, beginning prenatally, show increased male-typical toy, playmate, and activity preferences, and women with CAH show reduced heterosexual activity and interest (Hines, 2004). Normal variability in testosterone prenatally has also been related to normal variability in sex-linked toy, playmate, and activity preferences in healthy girls at the age of 3½ years (Hines et al., 2002). In addition, hormones have activational effects on sexuality; for instance, libido is increased by testosterone in adulthood (Alexander et al., 1997).

GONADAL HORMONES PRENATALLY AND HUMAN COGNITION (ORGANIZATIONAL INFLUENCES)

In the 1960s and 1970s, individuals exposed to high levels of androgens prenatally, because of CAH, were reported to have elevated IQ scores, a result interpreted to suggest that prenatal androgen exposure enhanced intelligence. However, subsequent research found that androgen-exposed participants, although scoring higher on intelligence tests than the general population, did not score higher than their relatives who did not have CAH or than control subjects who were carefully matched for background factors, such as family education and economic status (Hines, 2004). The high IQ scores appeared to have resulted from selection biases (i.e., individuals of high intelligence were more likely than those of low intelligence to participate in the research project). To mitigate selection biases, subsequent studies have typically enrolled relatives who were not exposed to hormones as control subjects or have carefully matched control subjects for background.

Because general intelligence does not show an appreciable sex difference, it is not surprising that it is not influenced by gonadal hormones. Abilities that show sex differences, such as mental rotations, spatial perception, mathematical ability, and verbal fluency, are more likely candidates. Studies investigating verbal fluency and mathematical abilities have generally not supported organizational hormone influences. In addition, although some researchers have concluded that prenatal exposure to androgens, particularly testosterone, enhances spatial abilities in females, this conclusion does not have a sound empirical basis. Of seven published studies assessing spatial abilities in females with CAH (see Table 7.1), only three have reported enhancement. Failure to find a link to testosterone in other studies does not appear to relate to the use of small samples or of measures that do not show sex differences. The three studies reporting enhanced spatial ability in females with CAH are not those with the largest samples. In fact, two of the three had the first and second smallest samples, and the two studies using the largest and second largest samples found no evidence of enhanced spatial ability; indeed one of them found impaired performance on a spatial perception task in women with CAH. Also, the tasks that show large sex differences are not the most likely to show the predicted link with androgen exposure. The three studies finding enhanced spatial ability in females with CAH used a total of five tasks—two mental rotations tasks and three spatial visualization tasks, a category of spatial ability that shows only a negligible sex difference. Only two of the seven studies used the 3-D mental rotations task that shows the largest sex difference. One of these found enhanced performance in females with CAH, but the second did not, even though it used a sample more than twice as large.

Studies relating normal variability in hormones prenatally or neonatally to cognitive performance also do not support an influence of testosterone. Neither testosterone in amniotic fluid nor testosterone in umbilical cord blood at birth shows the predicted relationship to subsequent cognitive performance (Finegan, Niccols, & Sitarenios, 1992; Jacklin, Wilcox, & Maccoby, 1988). In fact, more relationships have been observed in the direction opposite that predicted (e.g., testosterone relating negatively to spatial and numerical abilities) than in the anticipated direction. Although one study found a positive correlation between testosterone in amniotic fluid and speed of mental rotations in girls at age 7½ years, performance accuracy, which is the measure that shows a sex difference, did not relate to testosterone (Grimshaw, Sitarenios, & Finegan, 1995).

GONADAL HORMONES IN ADULTHOOD AND HUMAN COGNITION (ACTIVATIONAL INFLUENCES)

Early research on adult hormonal fluctuations and cognitive performance focused on the menstrual cycle and associated the pre- or perimenstrual phase with numerous deficits, including impaired general academic and exam performance. However, subsequent research did not support these claims.

More recently, the focus has shifted to cognitive abilities that show sex differences. Some reports have associated high estrogen cycle phases with impairment of abilities at which males excel and enhancement of abilities at

TABLE 7.1

Studies of Spatial Abilities in Individuals With Congenital Adrenal Hyperplasia

Source	Participants		Age (in years)	Findings and type of task
	CAH	Control		
Perlman (1973)	11F	11 (MT)	3–5	CAH F better (a).
Baker and Ehrhardt (1974)	13F, 8M	11F, 14M (RL)	4–26	No differences M or F (a, c).
McGuire, Ryan, and Omenn (1975)	15F, 16M	31 (MT)	5–30	No differences M or F (a).
Resnick, Berenbaum, Gottesman, and Bouchard (1986)	17F, 8M	13F, 14M (RL)	11–31	CAH F better on 3 (a, c, c), but no different on 2 (a, a).
				CAH M no differences.
Helleday, Bartfai, Ritzen, and Forsman (1994)	22F	22 (MT)	17–34	CAH F worse on 1 (b), but no different on 3 (a, a, c)[a].
Hampson, Rovet, and Altmann (1998)	7F, 5M	5F, 4M (RL)	8–12	CAH F better.
				CAH M worse on 1 (a).
Hines et al. (2003)	40F, 29M	29F, 30M (RL)	12–45	CAH F no different.
				CAH M worse on 1 (c).

Note. CAH = congenital adrenal hyperplasia; M = males; F = females; MT = matched; RL = relative; a = spatial visualization tasks; b = spatial perception task; c = mental rotations task. [a]For example, CAH F worse on 1 (b) but no different on 3 (a, a, c) means that CAH females performed worse than control participants on a spatial perception task (b) but no different from control participants on three other tasks, two of which were spatial visualization tasks (a, a) and one of which was a mental rotations task (c).

which females excel, with similar cognitive outcomes suggested to occur in women taking oral contraceptives containing estrogen (Hampson & Moffat, 2004; Kimura, 1999). However, details of results (e.g., specific measures that were affected, subgroups of participants who showed cognitive differences) suggest that factors other than estrogen could explain the findings (Hines, 2004). In addition, other researchers have not always been able to produce similar results (for reviews see Epting & Overman, 1998; Halari et al., 2005; Miles, Green, & Hines, in press).

A more rigorous approach to evaluating activational influences of hormones on cognition is to look at situations in which hormones are administered (e.g., as part of medical treatment). Results have not supported the hypotheses that estrogen treatment produces a more female-typical cognitive pattern or that androgen treatment produces a more male-typical pattern. Testosterone treatment has produced a variety of outcomes, including enhanced male-typical abilities, impaired male-typical abilities, and curvilinear relationships to male-typical abilities, as well as no effects on abilities (Halari et al., 2005; Hines, 2004; Miles et al., in press). Results for estrogen treatment (particularly of genetic males desiring to live as women) initially suggested an enhanced female-typical cognitive pattern, but several subsequent replication attempts did not produce similar results (Miles et al., in press). Treating adolescent girls with estrogen or adolescent boys with testosterone (for delayed puberty) also does not appear to influence performance on mental rotations tasks or other measures that show sex differences (Liben et al., 2002).

Effects of estrogen on cognition have also been evaluated by studying postmenopausal women. Some reports suggest that women on estrogen replacement show improved cognitive function (Kimura, 1999), but others do not (Barrett-Connor & Kritz-Silverstein, 1993). In addition, cognitive improvement, when seen, is not limited to tasks that show sex differences, and women who once took estrogen but no longer do so also show cognitive enhancement compared with those who never took estrogen. Both sets of findings suggest that selection biases could explain the cognitive advantage; women who are more cognitively able, or more highly educated, may be more likely to take estrogen at menopause. Consistent with this interpretation, some double-blind, placebo-controlled studies have found that estrogen treatment has no beneficial effect on cognition in postmenopausal women (see Miles et al., in press, for review).

Sex differences in cognition are smaller than sex differences in behaviors, such as childhood play interests or adult sexual orientation, for which influences of the early hormone environment have been clearly demonstrated. Thus, it could be suggested that more powerful methodology (e.g., larger samples, better measures) or hormone assessments at different times (e.g., neonatally) would reveal the expected effects. However, it cannot be assumed that such studies will reveal the anticipated effects, especially because some prior assumptions of hormonal or genetic influences have proved wrong.

SUMMARY AND CONCLUSION

In summary, sex differences in cognitive abilities have not been clearly linked to either organizational or activational effects of hormones. Similarly, genetic explanations have received little, if any, empirical support.

In addition, the suggestion that relatively small sex differences in certain very specific aspects of abilities explain the underrepresentation of women in science seems odd, given that scientific success depends on so much more than performance on tests of these abilities. For instance, math and science achievements in U.S. students lag far behind those of students in many other countries generally, and these national differences are larger than are sex differences. Describing the differences, Schmidt (1998) noted that "our best students in mathematics and science are simply not 'world class.' Even the very small percentage of students taking Advanced Placement courses are not among the world's best" (p. 2). This relatively poor performance among U.S. students, both male and female, has been noted for decades but has not led to suggestions that U.S. scientists are likely to be underrepresented among the top ranks.

Although born and educated in the United States, I have spent the past 11 years living and working in Great Britain, and I have noted differences in attitudes to the underperformance of girls versus boys in math and science. In Great Britain, girls routinely outperform boys on tests taken at the end of secondary school and used for university entry, including mathematics and science assessments. The reaction is not to conclude that boys are unlikely to succeed as scientists, but instead to wonder what needs to be done to boost boys' performance. A similar approach in the United States would not assume that sex equity in science is hopeless, but rather would devise ways to encourage girls to do better in areas in which they are underperforming.

Others who have studied the causes of sex segregation in occupations have concluded that the major determinants are economic and political, not genetic and hormonal. In all sex-segregated societies, males dominate occupations with higher prestige, power, and income, although the exact occupations involved can differ from one society to another. Historical shifts in our own society away from male dominance in fields such as teaching, secretarial work, and, more recently, medicine, can be more easily explained by changes in prestige, power, and income than by changes in hormones or genes.

Expectations and beliefs, as well as hormones and genes, influence the brain and behavior. A major contribution of psychology has been to show that expectations alone can cause dramatic changes in behavior. Placebo effects are one example. When led to believe that they are receiving a treatment (e.g., a pill) that will cause a certain effect, people will often show the expected effect, even though the pill is made only of sugar. Within the academic setting, Rosenthal and Jacobson (1968) demonstrated similar effects when teachers were told that certain students (selected at random) were about to blossom academically. Months later, not only were these randomly selected students rated by teachers as performing better than the other students, but they also performed better on objective measures of intelligence.

In the area of math achievement in particular, press reports of female disadvantage have been found to cause parents of girls to reduce expectations that their daughters will perform well in math (Jacobs & Eccles, 1985). These parental expectations, whether based in fact or not, could produce reduced performance in daughters, similar to the findings of Rosenthal and Jacobson (1968). They also could reduce the expectations of the daughters themselves, because parental expectations have been found to influence the expectations of their children (Tiedemann, 2000). In addition, individual expectations have been linked to math performance, as well as to interest among women in math and science (Correll, 2001; Frome & Eccles, 1998; Spencer, Steele, & Quinn, 1999). These results suggest that reports that innate factors, such as genes or hormones, limit girls' abilities in certain areas, even when incorrect, could well be self-fulfilling prophecies. Thus, sex differences in cognition need not cause a shortage of women in science, but they can if we let them.

REFERENCES

Alexander, G. M., Swerdloff, R. S., Wang, C., Davidson, T., McDonald, V., Steiner, B., et al. (1997). Androgen–behavior correlations in hypogonadal men and eugonadal men. *Hormones and Behavior, 31*, 110–119.

Baker, S. W., & Ehrhardt, A. A. (1974). Prenatal androgen, intelligence and cognitive sex differences. In R. C. Friedman, R. N. Richart, & R. L. Vande Wiele (Eds.), *Sex differences in behavior* (pp. 53–76). New York: Wiley.

Barrett-Connor, E., & Kritz-Silverstein, D. (1993). Estrogen replacement therapy and cognitive function in older women. *Journal of the American Medical Association, 269*, 2637–2641.

Cohen, J. (1988). *Statistical power analysis for the behavioral sciences* (2nd ed.). Hillsdale, NJ: Erlbaum.

Correll, S. J. (2001). Gender and the career choice process: The role of biased self-assessments. *American Journal of Sociology, 106*, 1691–1730.

Epting, L. K., & Overman, W. H. (1998). Sex sensitive tasks in men and women: A search for performance fluctuations across the menstrual cycle. *Behavioral Neuroscience, 112*, 1304–1317.

Finegan, J. K., Niccols, G. A., & Sitarenios, G. (1992). Relations between prenatal testosterone levels and cognitive abilities at 4 years. *Developmental Psychology, 28,* 1075–1089.

Frome, P., & Eccles, J. S. (1998). Parental effects on adolescents' academic self-perceptions and interests. *Journal of Personality and Social Psychology, 74,* 435–452.

Gould, S. J. (1981). *The mismeasure of man.* New York: Norton.

Goy, R. W., & McEwen, B. S. (1980). *Sexual differentiation of the brain.* Cambridge, MA: MIT Press.

Grimshaw, G. M., Sitarenios, G., & Finegan, J. K. (1995). Mental rotation at 7 years: Relations with prenatal testosterone levels and spatial play experiences. *Brain and Cognition, 29,* 85–100.

Halari, R., Hines, M., Kumari, V., Mehrotra, R., Wheeler, M., Ng, V., et al. (2005). Sex differences and individual differences in cognitive performance and their relationship to endogenous gonadal hormones and gonadotropins. *Behavioral Neuroscience, 119,* 104–117.

Hampson, E., & Moffat, S. D. (2004). The psychobiology of gender: Cognitive effects of reproductive hormones in the adult nervous system. In A. H. Eagly, A. Beall, & R. J. Sternberg (Eds.), *The psychology of gender* (2nd ed., pp. 38–64). New York: Guilford Press.

Hampson, E., Rovet, J. F., & Altmann, D. (1998). Spatial reasoning in children with congenital adrenal hyperplasia due to 21-hydroxylase deficiency. *Developmental Neuropsychology, 14,* 299–320.

Helleday, J., Bartfai, A., Ritzen, E. M., & Forsman, M. (1994). General intelligence and cognitive profile in women with congenital adrenal hyperplasia (CAH). *Psychoneuroendocrinology, 19,* 343–356.

Hines, M. (2004). *Brain gender.* New York: Oxford University Press.

Hines, M., Fane, B. A., Pasterski, V. L., Mathews, G. A., Conway, G. S., & Brook, C. (2003). Spatial abilities following prenatal androgen abnormality: Targeting and mental rotations performance in individuals with congenital adrenal hyperplasia (CAH). *Psychoneuroendocrinology, 28,* 1010–1026.

Hines, M., Golombok, S., Rust, J., Johnston, K., Golding, J., & the ALSPAC Study Team. (2002). Testosterone during pregnancy and childhood gender role behavior: A longitudinal population study. *Child Development, 73,* 1678–1687.

Hyde, J. S., Fennema, E., & Lamon, S. J. (1990). Gender differences in mathematics performance: A meta-analysis. *Psychological Bulletin, 107,* 139–155.

Hyde, J. S., & Linn, M. C. (1988). Gender differences in verbal ability: A meta-analysis. *Psychological Bulletin, 104,* 53–69.

Jacklin, C. N., Wilcox, K. T., & Maccoby, E. E. (1988). Neonatal sex-steroid hormones and cognitive abilities at six years. *Developmental Psychobiology, 21,* 567–574.

Jacobs, J. E., & Eccles, J. S. (1985). Gender differences in math ability: The impact of media reports on parents. *Educational Research, 14,* 20–25.

Kimura, D. (1999). *Sex and cognition*. Cambridge, MA: MIT Press.

Liben, L. S., Susman, E. J., Finkelstein, J. W., Chinchilli, V. M., Kunselman, S., Schwab, J., et al. (2002). The effects of sex steroids on spatial performance: A review and an experimental clinical investigation. *Developmental Psychology, 38,* 236–253.

Linn, M. C., & Petersen, A. C. (1985). Emergence and characterization of sex differences in spatial ability: A meta-analysis. *Child Development, 56,* 1479–1498.

Loehlin, J. C. (2000). Group differences in intelligence. In R. J. Sternberg (Ed.), *Handbook of intelligence* (pp. 176–193). New York: Cambridge University Press.

Lynn, R. (1999). Sex differences in intelligence and brain size: A developmental theory. *Intelligence, 27,* 1–12.

Maccoby, E. E., & Jacklin, C. N. (1974). *The psychology of sex differences*. Stanford, CA: Stanford University Press.

McGuire, L. S., Ryan, K. O., & Omenn, G. S. (1975). Congenital adrenal hyperplasia: II. Cognitive and behavioral studies. *Behavior Genetics, 5,* 175–188.

Miles, C., Green, R., & Hines, M. (in press). Estrogen treatment effects on cognition, memory and mood in male-to-female transsexuals. *Hormones and Behavior.*

Perlman, S. M. (1973). Cognitive abilities of children with hormone abnormalities: Screening by psychoeducational tests. *Journal of Learning Disabilities, 6,* 21–29.

Resnick, S. M., Berenbaum, S. A., Gottesman, I. I., & Bouchard, T. (1986). Early hormonal influences on cognitive functioning in congenital adrenal hyperplasia. *Developmental Psychology, 22,* 191–198.

Rosenthal, R., & Jacobson, L. (1968). Pygmalion in the classroom. New York: Holt, Rinehart & Winston.

Schmidt, W. H. (1998). *Are there surprises in the TIMSS twelfth grade results?* [Press release]. East Lansing: Michigan State University.

Spencer, S. J., Steele, C. M., & Quinn, D. M. (1999). Stereotype threat and women's math performance. *Journal of Experimental Social Psychology, 35,* 4–28.

Tiedemann, J. (2000). Parents' gender stereotypes and teachers' beliefs as predictors of children's concept of their mathematical ability in elementary school. *Journal of Educational Psychology, 92,* 144–151.

Voyer, D., Voyer, S., & Bryden, M. P. (1995). Magnitude of sex differences in spatial abilities: A meta-analysis and consideration of critical variables. *Psychological Bulletin, 117,* 250–270.

8

BRAINS, BIAS, AND BIOLOGY: FOLLOW THE DATA

RICHARD J. HAIER

Considerable evidence, reviewed in this volume, indicates that men and women show some differences in their respective patterns of cognitive strengths and weaknesses. Any such group differences are statistical in nature and never predict anything about a specific person's abilities. Most of the differences are rather small and may have no practical consequences. Nonetheless, the nature of these group differences may provide some insights into the origin of rather substantial cognitive differences among individuals. Some of my research addresses whether such differences reflect, at least in part, biological differences (genetic or not) in brain structure and function.

I was drawn recently into commenting on the subject of male versus female differences in cognition. My most recent research study concerning sex differences in brain structure as related to intelligence was published about the same time that Lawrence Summers, president of Harvard University, made controversial remarks about difficulties faced by women in science careers. From reading a transcript of his remarks, I believe he summarized the current situation reasonably well. As he noted, women face serious sex discrimination and child-rearing or family pressures. He also wondered whether women could be underrepresented at the very highest end of talent in science, especially fields that require advanced mathematics like engineering and physics, not only because of cultural bias, workplace discrimination, and unequal family pressures, but also because of possible biological or genetic reasons. This last remark caused consternation and bitter recriminations on the part of those who inferred he was proposing the biological or genetic inferiority of women.

For most researchers, this was an unfair inference, but not an uncommon one. Biological and genetic variables influence behavior, cognition, and the brain. No controversy exists about the importance of understanding how these influences work for Alzheimer's disease, schizophrenia, mental retardation, and a host of other serious problems. Controversy is more apt to arise when biology and genetics are studied as they influence behavior and cognition in the absence of neurological problems, as they surely do. In the case of Professor Summers's remarks, the scientific controversy is whether the brain parameters necessary for the most advanced mathematical reasoning may be unequally distributed between men and women, and if so, why. There are no definitive answers yet to either part of this straightforward question, but there are interesting scientific data to consider. None of the data are correctly interpreted as showing the superiority or inferiority of one group or another.

Many studies show male versus female differences in cognitive abilities. Many other studies show male versus female differences in brain structure and function. Do such brain differences have connection to cognitive differences or anything else? So what if one group, on average, has a bigger brain, or more gray matter, or more white matter, or more connections between right and left hemispheres? Do any of these brain differences correlate to differences in general intelligence, to specific cognitive abilities, to vulnerability to dementia, or to other neurological diseases? Do any brain differences predict which individual will respond to a psychotropic drug or which stroke victim will have a full recovery? These are important and serious questions worth researching. They have little to do with popular psychology questions like why men allegedly do not ever ask for or read directions.

Earlier research conducted by my colleagues and me used functional brain imaging with positron emission tomography (PET) to investigate whether men and women, matched for SAT—Mathematics (SAT–M) scores, activated the same or different brain areas during a test of mathematical reasoning (see Haier & Benbow, 1995). There were 22 men and 22 women of college age; half of each group were admitted to college with scores over 700 on SAT–M (a perfect score is 800), the other half scored in the average range with a score about 500. Each person completed a new SAT–M test while undergoing the PET scan procedure. The results showed that the harder the temporal lobes were working in the men, the better their score. This relationship between temporal lobe activation and performance on the math reasoning test was not found in the women, nor was activity in any other brain area related to the women's SAT–M score. Here was a clear sex difference in brain function, although it was unsatisfying that no regional activity related to math performance in the women. How such highly able women did the math reasoning without activating (or deactivating) specific brain areas was a mystery.

Our recent research used magnetic resonance imaging (MRI) and asked whether brain structure, especially the amount of gray and white matter in different brain areas, was related to general intelligence, as determined by standard IQ tests in normal volunteers ($N = 47$). Apparently, it is (see Haier, Jung, Yeo, Head, & Alkire, 2004). There are structures distributed throughout the brain where the amount of gray matter or white matter predicts IQ score. Specific areas associated with language in the frontal and parietal lobes seem especially important. Other researchers have shown that the volume of these same brain areas appears to be under genetic control.

Because general intelligence does not differ between men and women, we had no reason to expect sex differences in the brain structures related to IQ. However, we were wrong. When we reanalyzed our MRI data separately for men and women, we found completely different brain areas correlated to IQ (the men and women in these samples were matched on IQ). The amount of gray and white matter in the frontal areas seems more important in the women; the gray matter in the parietal areas seems more important in the men (Haier, Jung, Yeo, Head, & Alkire, 2005). This apparent sex difference is the finding that received significant public attention following Professor Summers's remarks. If this difference proves to be correct after independent replication, it could be concluded that men and women achieve the same general cognitive capability using different brain architectures.

Why is the notion of differential brain architecture important? It would mean that not all brains work the same way to achieve the same result, negating a principal assumption of traditional cognitive psychology. It would put the concept of individual differences in the center of human brain research and refocus attention on questions like why do some individuals learn, memorize, and reason better than others; it also may help explain why some pain drugs work better in women than in men. It also could have important clinical implications for understanding individual and group differences in the impact of brain damage and for cognitive rehabilitation strategies. For example, given our MRI and intelligence findings, a stroke or tumor in the frontal lobe may have different cognitive consequences for women than for men. For another example, Alzheimer's disease reaches the frontal lobes later than other areas, so it may be possible that women have the disease longer before any cognitive symptoms are apparent. This could lead to women receiving the diagnosis later in the disease process, delaying efficacious treatment with drugs now under development.

Questions about group differences often raise difficult social issues because they focus on how people differ rather than on how all humans are the same. It is still fashionable to ascribe behavioral and cognitive differences among people entirely to cultural and environmental differences, especially in childhood. This remains true even though many twin or sibling studies in behavioral genetics consistently show zero contribution of shared environment to variance in psychological variables like personality and intelligence. Because such variables do show moderate to high heritability, we know there is a genetic component of importance. Because genes always work through biology, there must be some biological basis for intelligence and personality. We also know that genes and their expression must be influenced by nongenetic factors. For example, identical twins are not nearly 100% concordant for schizophrenia (i.e., if one twin is schizophrenic, the other has only a 50% chance although both are genetically identical). Even more striking is the relatively recent revelation from the Human Genome Project that there are fewer than 25,000 human genes, a fraction of the number originally predicted. Because there are over 2 million gene products, this means each gene can express itself in a thousand different ways. The mechanisms that lead a gene to one expression or to any of a thousand others are not understood (Silverman, 2004). Whatever these mechanisms turn out to be, some of them are likely to be influenced, at least in part, by nongenetic factors including social and cultural ones.

What does all this have to do with women in science? Was the president of Harvard correct to wonder whether there are more men than women with advanced mathematical reasoning ability because of biological or genetic reasons? It is a question I have wondered about since my first year of graduate school in psychology at Johns Hopkins University in 1971. Professor Julian Stanley was conducting the first search for mathematically talented junior high school students (a project still under way). I helped with the SAT–M testing of about 450 Baltimore area students recommended by their math or science teachers. The students were mostly 12 to 14 years old. Twenty-three scored over 650 on the SAT–M, a score at the 94th percentile for college-bound seniors. Forty-three boys scored higher than the highest scoring girl (her score was 600; Stanley, Keating, & Fox, 1974).

Long-term studies with hundreds of thousands of students do show more mathematically gifted males than females, and many possible explanatory variables have been examined (see Lubinski and Benbow, chap. 6, this volume; Camilla Persson Benbow was also a graduate student with Professor Stanley). The ratio of boys to girls from the original Hopkins testing is not constant over time nor is it constant across countries. The sample sizes at 0.1% are very small, so inconsistencies are to be expected. Nonetheless, it must be noted that any inconsistency in itself does not argue against a possible genetic component. There are more blue-eyed people in Iceland than in Tibet, but it would be wrong to conclude from this fact that blue eye color is not genetic. At this stage of research, it is not certain that genes play a role either in mathematical talent among individuals or in any sex differences in the size of the talent pool at this level of ability, but it is a fair and important question.

A role for genes is strongly suspected in rare cases of mathematical genius, especially those in which environmental and cultural variables discouraged achievement. For example, Srinivasa Ramanujan is regarded as one of the great math geniuses of all time (Kanigel, 1991). He was born in poverty in India in 1887. In school, he was mostly self-taught in mathematics from outdated books, but he created profoundly original proofs and conjectures, which came to the attention of the best mathematicians at Cambridge University. As a young man, Srinivasa was brought to Cambridge, where he produced one astounding theorem after another until he died at age 32. Emilie de Breteuil, Marquise du Chatelet, however, was born into French aristocracy in 1706 (Osen, 1974). She had excellent tutors and a fine education. She led a life of wealth and privilege, spending much time with Voltaire and other intellectuals. She also showed an early genius for mathematics. It is reported that she could divide a nine-digit number by another nine-digit number in her head. Emilie conducted scientific experiments on motion and speed and published research works despite intense cultural bias against women pursuing such intellectual activities. Before she died at age 43, she demonstrated mathematically that velocity squared was the appropriate measure of kinetic energy; this is regarded as a major contribution and an important basis for Albert Einstein's most famous equation more than 150 years later.

Of course, two cases of exceptional genius tell us nothing about the size of the pool of mathematical talent let alone any differences in the pool between men and woman. They do suggest, however, that despite environmental deprivations or cultural limitations, genius can emerge. Did these two individuals share similar, but rare, neurotransmitter activity, brain structure connections, unusually large brain areas in the frontal and parietal regions, or other cerebral anomalies that resulted in their mathematical genius? To what extent did unknown genetic factors contribute to these brains?

Over time, modern societies seek to minimize cultural and environmental disadvantages and overt discrimination as best as they can to promote a level playing field for everyone. As these social goals are achieved, the differences that remain among people, especially at the highest levels of talent, will be attributed more and more to genetic factors. It is commonly believed that anything genetic is fixed in stone and cannot be easily altered, at least not as easily as any environmental variable. In the 21st century, just the opposite may be true. As we learn more about genetic engineering and the mechanisms for multiple gene expressions, there may be ways to influence brain development, growth, and function as they relate to specific and to general mental abilities.

So now we come to speculation beyond the current scientific findings but well within the imagination of scientists and writers of science fiction. Brain imaging research on intelligence is identifying the specific areas important for reasoning ability (Jung & Haier, 2006). These include frontal areas and parietal areas, especially Brodmann area (BA) 39 in the left parietal lobe (Brodmann areas are a standard numerical nomenclature for regions of the cortex). Compared with normal control subjects, BA39 was 15% bigger in Einstein's brain studied after his death. Gray matter volumes in the frontal and parietal lobes also appear to be under genetic control. If mathematical and scientific thinking is best accomplished in individuals with bigger BA39s and more connections between BA39 and the frontal lobes, what do we need to know before we can increase the volume of BA39 and its connections in any person who wishes to do so? Is it just a matter of finding the right brain growth factors, biological or nonbiological? With the right tweaking of the brain, can an average math student become gifted? Can a gifted student become a math genius? If such factors exist, are they the same in men and women? How could such factors be stimulated and controlled? What would be the best techniques—brain surgery, drugs, diet, listening to Mozart?

This is an important direction in neuroscience research, and the results likely will be as controversial as any issue involving the application of new biological or neuroscience knowledge, perhaps more so because the brain is involved. It will build on current work on drugs being developed for Alzheimer's disease by trying such medications in nonimpaired people to increase their memory ability from normal to extranormal. It will build on research implanting electrical stimulators, or stem cells, into specific brain areas to alleviate the symptoms of Parkinson's disease and other brain disorders. Can such electrical or stem cell stimulation to other brain areas improve cognition even in people who do not have any impairment or disorder? If any of this comes to pass, such techniques may work better in men or in women, depending on any differences in the salient brain architectures underlying cognition. If there is a disparity in any cognitive ability like talent for mathematical reasoning, which likely is based in part on biological and genetic aspects of the brain, neuroscience research someday may help minimize the disparity. Is this any less desirable or any more controversial than minimizing cognitive disparities by applying any social or cultural research results?

The hardest part of science is going wherever the data take you. Nature is the way it is, no matter how we think it should be. Yet, once empirical facts become known through scientific inquiry, there is always the possibility of changing nature. This is certainly true in biology. Every time you visit a physician, it is with the expectation that broken biology can be fixed, even if there is a genetic basis for the problem, and even if the fix involves changes in diet or exercise rather than medication. For any scientific question, especially about the brain and differences in cognitive performance, the data may point in uncomfortable directions, but gathering data from all potentially relevant sources is required for understanding the depth of the problem.

Research often takes unexpected turns not foretold by a priori hypotheses or by popular expectations. Given the pressures of funding, publication, and peer acceptance, such turns can be unwanted intrusions, or they can be thought-provoking opportunities that lead to original discoveries and controversial applications. The challenge is to follow where the data lead, always cognizant of Orwellian fears and prejudiced misuse of knowledge balanced by the prospects of alleviating suffering from disorders and enhancing the quality of life for everyone. Along the way, controversy can only escalate as we constantly test new knowledge against old and comfortable ideas. This is the way science works and the way our culture evolves.

REFERENCES

Haier, R. J., & Benbow, C. P. (1995). Sex differences and lateralization in temporal lobe glucose metabolism during mathematical reasoning. *Developmental Neuropsychology, 11,* 405–414.

Haier, R. J., Jung, R. E., Yeo, R. A., Head, K., & Alkire, M. T. (2004). Structural brain variation and general intelligence. *NeuroImage, 23,* 425–433.

Haier, R. J., Jung, R. E., Yeo, R. A., Head, K., & Alkire, M. T. (2005). The neuroanatomy of general intelligence: Sex matters. *NeuroImage, 25,* 320–327.

Jung, R. E., & Haier, R. J. (2006). *The parieto-frontal integration theory (P–FIT) of intelligence: Converging neuroimaging evidence.* Manuscript submitted for publication.

Kanigel, R. (1991). *The man who knew infinity: A life of the genius, Ramanujan.* New York: Scribner's.

Osen, L. M. (1974). *Women in mathematics.* Cambridge, MA: MIT Press.

Silverman, P. H. (2004). Rethinking genetic determinism. *The Scientist, 18*(10), 32–33.

Stanley, J., Keating, D. P., & Fox, L. H. (1974). *Mathematical talent: Discovery, description, and development.* Baltimore: Johns Hopkins University Press.

9

SCIENCE, SEX, AND GOOD SENSE: WHY WOMEN ARE UNDERREPRESENTED IN SOME AREAS OF SCIENCE AND MATH

DIANE F. HALPERN

It is a deceptively simple question: Is the underrepresentation of women in the sciences and math caused by sex differences in cognitive abilities? Of course, the real question is not neutral—it is about a presumed deficiency in women: Are there too few women with the cognitive abilities that are needed for careers in science and math? There are several assumptions embedded in this question, including the assumption that there is a direct relationship between cognitive ability and career choice and success. Because these essays are relatively short, I address the main assumptions separately, consider what is missing from the way the question is asked, and put the "pieces" together in a summary section in which a clear statement of the answer is provided. There are multiple, circuitous paths, some with detours and dead-ends leading from ability to career; the topic is complex, and the landscape is fraught with political minefields. The underlying question is about a broad range of important societal issues and how we think about biology, sex role socialization, education, and the roles and future roles of women and men in society.

ARE WOMEN REALLY UNREPRESENTED
IN SCIENCE AND MATH?

Many years ago I wrote a paper about cognitive sex differences subtitled, "What You See Depends on Where You Look" (Halpern, 1989). This homily could be repeated every time someone reports that there is or is not a sex difference with regard to some variable. Consider the sciences: There are many different kinds of sciences, and the graduation rates of women vary widely among them. Women are obtaining 50% of the MD degrees from medical schools, almost 75% of the VMDs from veterinary schools, and 44% of PhDs in biology and life sciences. It is clear that women are succeeding in some sciences and thus must have the cognitive ability to learn and succeed in science. So, it is not a lack of cognitive ability that is responsible for their underrepresentation, and there are careers in the sciences and math that can accommodate a wide range of abilities. Women are not achieving equally in all areas of science. They are obtaining only 29% of the PhDs in mathematics, 17% in engineering, and 22% in computer and information sciences.

What is also interesting as we consider women's success in science and math is the inherent value system in the way the question is asked. Why has no one asked, "Do men have the cognitive ability to achieve in areas in which they are underrepresented?" Men obtain only 32% of the PhDs in psychology, 37% of the PhDs in health sciences, and 34% of the PhDs in education. The distinction in this sex-differentiated list of sciences has been described as biological and health sciences for women versus inorganic sciences for men. Yet women have always tended to choose helping careers such as teaching, education, social work, and nursing. We can now add to that list veterinary medicine, medicine, and law, in which women are also graduating at approximately 50%. These are all careers that are people oriented, and some are science oriented as well. Men are overrepresented in mechanical and technical careers such as refrigerator and computer repair, accounting and business, physics, chemistry, and mechanical engineering. When careers are viewed along this dimension, the grouping indicates that women and men differ in the relative strength of their interest on the "people versus things" dimension of individual differences scale (Lippa, 1998), with more women toward the "people" end and more men toward the "things" or mechanical end.

Ackerman and his colleagues (Ackerman, Bowen, Beier, & Kanfer, 2001) have been particularly interested in the way trait complexes, which are combinations of interest and personality variables, combine with cognitive abilities to influence achievement. Individuals engage in activities that reflect their interests and their personalities, and in this way, they enhance the knowledge they need to achieve in a field. Interests are manifested in other ways, such as choice of major in college and career choices. Researchers in this area generally conclude that sex differences in coursetaking patterns and test scores reflect differences in personalities and values. Thus, career choices reflect activities that stem from abilities but include more than cognitive abilities.

ARE THERE SEX DIFFERENCES IN COGNITIVE ABILITIES?

On average, there are some cognitive tests and tasks that show differences between girls and boys and between women and men. It is important to understand that most people prefer to talk about similarities and are afraid that any discussion of differences will be misinterpreted and used in ways that are misogynistic. The fear of misuse of research findings is understandable, given the long history of women's oppression and current events that include the abortion of female fetuses and female infanticide in many places in the world, the documented preference for sons in the United States, and numerous instances of women's rights being lost as some countries justify extreme interpretations of religious texts. Yet the data on cognitive sex differences do not show a smarter or better sex, only that there are some differences; nor do they impute a necessary origin for the differences (references for these assertions are in Halpern, 2000, 2004).

Although most people have a preference for environmental explanations for average differences between the sexes on cognitive tests or tasks, we cannot separate the environmental effects from the biological ones. Learning is both an environmental and a biological phenomenon, so when and how quickly and how well something is learned will depend on prior learning, opportunities for prior learning, interest, and many more variables. Figure 9.1 depicts a biopsychosocial model (also known as a psychobiosocial model) in which individuals seek learning experiences from their environment. These experiences then influence their interests, which then directs their further learning and alters the underlying neural structures in their brain, which in turn directs additional learning.

Standardized intelligence tests have been normed to show no overall difference between males and females, and other researchers who have looked at tests that were not specifically normed to show no overall sex difference have concluded that on average, there is no difference in intelligence between females and males. In other words, overall, the sexes are equally "smart."

Now, given these caveats, consider some of these sex differences. In general, girls get higher grades in school, so most school-related measures of achievement, including tests closely related to material learned in school, favor girls. Girls develop at a faster rate than boys, have many fewer behavioral problems, and are more likely to graduate from high school. Women comprise a substantial majority of college enrollments in the United States and many other countries. (Women in the United States have received more college degrees than men every year since 1982, with the gap widening every

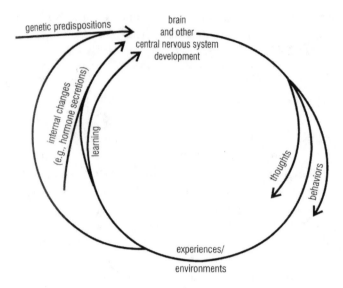

Figure 9.1. A biopsychosocial model as a framework for understanding cognitive sex differences. It replaces the older idea of nature versus nurture with a circle that shows the way biological and psychosocial variables exert mutual influences on each other. Copyright 2000 by D. F. Halpern. Printed with permission.

year; for example, among women between 25 and 34 years old, 33% have completed college compared with 29% of men.) Despite these "successes," women score significantly lower on many (not all) standardized tests of science and mathematics, when the tests are not closely related to material as it has been taught in school. This discrepancy has led to calls of bias from just about every portion of the political spectrum, leading to at least two types of questions (a) are teachers and schools biased against boys? (b) are standardized tests biased against girls?

In a report published by the U.S. Department of Education titled *Trends in Educational Equity of Girls and Women*, the data on reading and writing achievement are described this way:

> Females have consistently outperformed males in writing achievement at the 4th, 8th, and 11th grade levels between 1988 and 1996. Differences in male and female writing achievement were relatively large. The writing scores of female 8th graders were comparable with those of 11th grade males. (Bae, Choy, Geddes, Sable, & Snyder, 2000, p. 18)

Compare their conclusion with the meta-analytic review of the research literature by Hedges and Nowell (1995), who reported that "the large sex differences in writing . . . are alarming" (p. 41). The advantage for girls on tests of writing is large and robust. Thus, one way to think about the difference in test scores is whether the assessment contains a lengthy written response, which is likely to benefit girls. The ability to communicate well in writing is a necessary skill in the sciences and math professions, even if it is less valued in science and math education. Scientists and mathematicians need to write grants, write up research results, communicate their work with peers, and so on. Written communication is a necessary skill for success in any field and should be an advantage that is not paying off for women in science and math careers.

Visual–spatial abilities are also essential in the sciences and math. These are abilities such as being able to visualize a complex image and maintain it in memory while "working" on it. The sizes of the differences between females and males (effect sizes) for tasks that require the generation of an image and maintaining it in memory while "working" on it vary. Depending on the complexity of the image to be generated and the nature of the task, the differences can be fairly large and favor males (in technical terms they range between .63 and .77 standard deviations; see references in Halpern, 2000). Mental rotation tasks, which require the maintenance of three-dimensional figural information in working memory while simultaneously transforming information, show very large sex differences in favor of males—they are among the largest differences found in the psychological literature (somewhere between 0.9 to 1.0 standard deviations). These differences are so large that no statistical tests are needed to determine group differences.

The lone visual–spatial task that shows an advantage for females is memory for objects and their locations, a finding that occurs consistently, even with different testing paradigms or setups. In fact, females score higher than males on many measures that involve memory, including memory for personal information, in general, or memory that includes location information (episodic memory), associative memory (recalling words in pairs), and a composite of memory tests, just to name a few examples.

Male scores tend to show greater variability (more "spread") than those of females (Hedges & Nowell, 1995), with a higher percentage of males than females in both the high-ability and low-ability tails of the math and visual–spatial distributions. As a consequence, when combined with mean differences favoring males, the *tail ratio*, or ratio of males to females at extremely high-ability levels, such as the top 1% to 0.5%, strongly favors males, despite more modest differences in the middle of the distribution. Several researchers have argued that the excess of males at the very high end of the abilities distributions for mathematics can account for the underrepresentation of females in physical sciences and math careers. However, there is a lack of females at all abilities ranges in physical sciences and math, not just at the highest abilities range, and there must be many males in physical sciences and math careers who are not in the highest ability ranges because, by definition, only a very small percentage of the population are in this range.

A catalogue of sex differences in cognitive abilities provides no information about the reasons for these differences. There are many laws of cognitive psychology that are as certain as the law of gravity. Here is one of them: Except for those individuals who are so profoundly mentally retarded that they will not benefit from education, all cognitive abilities will improve with learning and practice, which is the reason why we have schools. Everyone improves in each area with education. None of these differences is immutable.

IS THERE GOOD EVIDENCE THAT SEX STEROID HORMONES AFFECT COGNITIVE ABILITIES?

Mention anything about sex differences, and someone will say it is all in the hormones. Yet is there good evidence that sex steroid hormones influence cognitive abilities in a meaningful way? Sex steroid hormones are produced by the gonads (testes in men and ovaries in women) and, to a lesser degree, the adrenal glands. The hormones include androgens, such as testosterone; estrogens, such as estradiol; and progestins, such as progesterone. All categories of hormones are produced in males and females; however, androgens predominate in males, and estrogens and progestins predominate in females. If hormones alter cognitive abilities, they could do so directly, by influencing the development or activity of specialized neural areas, or indirectly, through an effect on related cognitive skills or on life experiences. In humans, for example, hormones could alter the development of a cognitive ability by predisposing a person to seek out (or avoid) certain activities, such as playing with certain toys like trucks or dolls that foster the development of one type of cognitive ability over another. Studies of girls who were exposed to abnormally high levels of prenatal androgens suggest that this may be one possible mechanism for sex-differentiated cognitive development. However, it is important to keep in mind the fact that hormones do not produce fixed, preprogrammed effects; they interact in a complex manner with environmental and experiential factors, so they can never be considered independent of a person's learning environment.

Recent work has shown that there are lifetime cumulative effects of estrogen on a battery of cognitive tasks. Psychologists have computed lifetime exposure to estrogen for a sample of healthy older women and found that those women who had had greater exposure to estrogen (e.g., early age of menarche and late menopause) had higher scores on a battery of cognitive tasks than women with shorter exposures to estrogen. Although every conclusion about sex steroid hormones has been the object of intense debate among researchers, most will now agree that there are sex differences in the shape, and probably the volume, of a portion of the corpus callosum, a thick band of neural fibers that connects the two hemispheres of the brain, with females in general having a larger and more bulbous structure, which could have developed in response to prenatal hormones, postnatal hormones, experiences, or a combination of all or some of these.

The burgeoning field of hormone replacement therapies for men and women is providing evidence that hormones continue to be important in cognition throughout the life span, although the field is complex and rife with controversies. The best evidence for a beneficial effect is the effect of estrogen on verbal memory in old age. There are still many studies that have failed to find beneficial effects for hormone replacement in elderly women; however, there are a substantial number of studies that suggest that exogenous estrogen (pill, patch, cream, or other form) causes positive effects on the cognition of healthy older women and possibly for women in early stages of Alzheimer's disease. This conclusion is in accord with Sherwin's (1999) meta-analytic review of 16 prospective, placebo-controlled studies in humans, in which she concluded the following:

> Estrogen specifically maintains verbal memory in women and may prevent or forestall the deterioration in short- and long-term memory that occurs with normal aging. There is also evidence that estrogen decreases the incidence of Alzheimer disease or retards its onset or both. (p. 315)

The results of these studies and others provide a causal link between levels of adult hormones and sex-typical patterns of cognitive performance.

But the evidence is even stronger that experience alters brain structures and cognitive abilities. The distinction between biology and experience is hopelessly blurred when you consider that the architecture of the brain is also shaped by experience, so it cannot be assumed that differences in female and male brains result solely from hormonal action. The old nature–nurture dichotomy is surely wrong; nature and nurture are inseparable. A study that made front-page news in many major newspapers found that taxi cab drivers in London had enlarged portions of their right posterior hippocampus relative to a control group of adults whose employment required less use of spatial skills. The cab drivers showed a positive correlation between the size of the region of the hippocampus that is activated during recall of complex routes and the number of years they worked in this occupation. The finding that size of the hippocampus varied as a function of years spent driving taxis makes it likely that it was a lifetime of complex wayfinding that caused the brain structure used in certain visual–spatial tasks to increase in size, although other explanations also are possible.

What about explanations that are not about cognitive abilities? There are many possible explanations for the disproportional representation of women and men in different areas of science and math. One area of research has shown the importance of the unconscious effects of stereotypes on thought and performance, called *stereotype threat*, which is the idea that if students are being assessed in an area in which there is a negative stereotype about their group's ability, an unconscious "threat" will be activated and that threat will decrease performance. There is a large and old literature in psychology and education on self-fulfilling prophesies showing that experimenter or teacher expectations can unconsciously influence how people respond to situations. Medical researchers are well aware of these types of effects, which is why double-blind, placebo-controlled, crossover studies are the "gold standard" for medical research. Work by Steele and Aronson and others has extended these principles to explain how beliefs about the cognitive abilities of different groups can cause or contribute to group differences on tests of cognitive abilities. The idea is that by making the fact that the test-taker is female or male salient at the time a cognitive test is being administered, commonly held beliefs about the performance of females or males are activated. Test-takers are "threatened" by these beliefs out of the concern that they will conform to their group's negative stereotype. According to this theory, stereotype threat will only affect test performance when (a) the group membership is made salient, (b) the test that is being taken is relevant to one's group (e.g., the stereotype that females are not as good in mathematics as males), (c) test performance is important to the individuals taking the test, and (d) the test is at a level of difficulty that the additional burden of defending against a perceived threat would cause a performance decrement. There is still much we do not know about the parameters under which stereotype threat operates and why there are still so many controversies over situations in which it failed to operate under real world conditions. It is likely to be resolved in the next few years as more researchers tease out the relevant variables.

SO, DO SEX DIFFERENCES IN COGNITION EXPLAIN SEX DIFFERENCES IN SCIENCE, MATH, OR ANY OTHER CAREER OR JOB?

Jobs can be disproportionately male or female for many different reasons unrelated to cognitive ability. For example, it is likely that there are few female piano movers in New York City because women, on average, have less upper body strength than men, not because they lack the "smarts" for the job. Other highly sex-segregated jobs are more readily explained by the fact that women do the bulk of the caretaking responsibilities in society (child care, sick care, and elder care), and therefore often work "around" these responsibilities. As caretakers, many women need careers that offer more flexibility and more autonomy in how they schedule their work time, so they can care for children and others, such as disabled and older family members. Consider, for example, how this alternative hypothesis can be used to explain the underrepresentation of women in high-level science professions.

Academia is one of the few places in which young, talented professionals have to prove themselves at a young age to keep their job. If graduate school is followed by a post doc (as many in the sciences will do) and then 6 years at the assistant professor level, the young academic will be approximately 36 years old before applying for tenure (assuming everything has gone smoothly), and if denied tenure, she or he will be fired ("not renewed") and has to explain why tenure was denied to hiring committees at some less prestigious college. For women, tenure clocks and biological clocks run on the same time zone, and although maternal and paternal leaves are available at most universities, there are also subtle and not so subtle pressures not to take advantage of these leaves. The conditions of academic life are particularly difficult for any woman who has caregiving responsibilities such as child care, which is a more likely reason for the underrepresentation of women in academic science, with its additional requirements for laboratory hours, than the fewer number of women at the highest tails of math and science standardized tests. The fewer numbers of men with high scores on writing tests or with high grades in school were ignored, and the tenure system with its requirements to show excellence at a young age and insistence on full-time employment were not questioned as gender-fair workplaces, when the question about the underrepresentation of women in science was posed.

THE TAKE-HOME MESSAGE

The data presented here show that there is no evidence that one sex is "smarter" than the other, and the reason is not just because standardized multiscore intelligence tests were written that way. Women get higher grades in school and are graduating in numbers substantially higher than men in veterinary medicine, approximately equal to men in medicine, and close to equal numbers in life and health sciences. Men are represented in much higher numbers in the extreme tails of standardized tests of mathematics and have a much higher proportion of careers in the physical sciences, engineering, and mathematics. Women are underrepresented at all ability ranges from the physical sciences and math. These data do not support the notion that the differences in participation rates in science and math are caused by differences in cognitive abilities. Women are the caretakers in society. What is needed is a society in which sex roles are more equalized so that men share in child care and other caregiving, institutions are aligned with the needs of women and men throughout their life spans, and opportunities to develop a range of talents and interests are offered early in life and supported for all girls and boys as they grow and consider career options. Part-time and other flexible tenure track career options that vary as needed through the life span would be a giant step toward attracting and keeping high-ability women and men in the full range of science and math careers.

REFERENCES

Ackerman, P. L., Bowen, K. R., Beier, M. E., & Kanfer, R. (2001). Determinants of individual differences and gender differences in knowledge. *Journal of Educational Psychology, 93*, 797–825.

Bae, Y., Choy, S., Geddes, C., Sable, J., & Snyder, T. (2000). *Trends in educational equity of girls and women* (NCES 2000-030). Washington, DC: U.S. Department of Education, National Center for Education Statistics.

Halpern, D. F. (1989). The disappearance of cognitive gender differences: What you see depends on where you look. *American Psychologist, 44*, 1156–1158.

Halpern, D. F. (2000). *Sex differences in cognitive abilities* (3rd ed.). Mahwah, NJ: Erlbaum.

Halpern, D. F. (2004). A cognitive-process taxonomy for sex differences in cognitive abilities. *Current Directions in Psychological Science, 13*, 135–139.

Hedges, L. V., & Nowell, A. (1995, July 7). Sex differences in mental test scores, variability, and numbers of high-scoring individuals. *Science, 269*, 41–45.

Lippa, R. (1998). Gender-related individual differences and the structure of vocational interests of the people–things dimension. *Journal of Personality and Social Psychology, 74*, 996–1009.

Sherwin, B. B. (1999). Can estrogen keep you smart? Evidence from clinical studies. *Journal of Psychiatry and Neuroscience, 24*, 315–321.

10

WOMEN IN SCIENCE: GENDER SIMILARITIES IN ABILITIES AND SOCIOCULTURAL FORCES

JANET SHIBLEY HYDE

In the United States, women are underrepresented at the highest levels in the physical sciences but not in the biological sciences. In 2000, women earned 25% of the PhDs in the physical sciences, but they were near parity (43%) with men for PhDs in the biological sciences (National Science Foundation, 2002). In seeking to explain the underrepresentation of women in the physical sciences, we can ask three crucial questions: (a) Do women, compared with men, lack the abilities needed for success in the physical sciences? That is, is there a gender difference in these abilities favoring men? (b) If gender differences exist, are they the result of sociocultural factors? and (c) Depending on the answers to the first two questions, how can we close the gender gap in the physical sciences? This chapter addresses each question in turn.

Success in the physical sciences requires many abilities. Chief among them are mathematical, spatial, and verbal abilities, the first two for doing

Preparation of this chapter was supported in part by National Science Foundation Grant REC 0207109.

the science and the third for presenting one's work in scientific articles and at conferences. Researchers have amassed mountains of data on gender differences in mathematical, spatial, and verbal abilities. They have synthesized the results of all these studies using a statistical technique called *meta-analysis*. Therefore, before reviewing the evidence on gender differences in abilities, I provide a brief explanation of meta-analysis.

META-ANALYSIS

Meta-analysis is a statistical method for aggregating research findings across many studies of the same question (Hedges & Becker, 1986). It is ideal for synthesizing research on gender differences, an area in which often dozens or even hundreds of studies of a particular question have been conducted.

Crucial to meta-analysis is the concept of *effect size*, which measures the magnitude of the effect—in this case, the magnitude of the gender difference. In gender meta-analyses, the measure of effect size typically is d (Cohen, 1988):

$$d = \frac{M_M - M_F}{s_w}$$

where M_M is the mean score for males, M_F is the mean score for females, and s_w is the within-sex standard deviation. That is, d measures how far apart the male and female means are, in standardized units. In meta-analysis, the effect sizes computed from all individual studies are then averaged to obtain an overall effect size reflecting the magnitude of gender differences across all studies. Here I follow the convention that negative values of d indicate that females scored higher and positive values of d indicate that males scored higher. Although there is some disagreement among experts, a general guide is that an effect size d of 0.20 is a small difference, a d of 0.50 is moderate, and a d of 0.80 is a large difference (Cohen, 1988).

Meta-analyses generally proceed in three steps. First, the researchers locate all studies on the topic being reviewed, typically using databases such as PsycINFO and carefully chosen search terms. Second, statistics are extracted from each report, and an effect size is computed for each study. Third, an average of the effect sizes is computed to obtain an overall assessment of the direction and magnitude of the gender difference when all studies are combined.

Conclusions based on meta-analyses are almost always more powerful than conclusions based on an individual study, for two reasons. First, because meta-analysis aggregates over numerous studies, a meta-analysis typically represents the testing of tens of thousands—sometimes even millions—of participants. As such, the results should be far more reliable than those from any

TABLE 10.1
Magnitude of Gender Differences in Mathematics Performance as a
Function of Age and Cognitive Level of the Test

Age group (years)	Cognitive level		
	Computation	Concepts	Problem solving
5–10	−0.20	−0.02	0.00
11–14	−0.22	−0.06	−0.02
15–18	0.00	0.07	0.29
19–25	NA	NA	0.32

Note. From "Gender Differences in Mathematics Performance: A Meta-Analysis," by J. S. Hyde, E. Fennema, and S. J. Lamon, 1990, *Psychological Bulletin, 107,* p. 148. Copyright 1990 by the American Psychological Association.

individual study. Second, findings from gender differences research are notoriously inconsistent across studies. For example, in the meta-analysis of gender differences in mathematics performance discussed later in this chapter, 51% of the studies showed males scoring higher, 6% showed exactly no difference between males and females, and 43% showed females scoring higher (Hyde, Fennema, & Lamon, 1990). This makes it very easy to find a single study that supports one's prejudices. Meta-analysis overcomes this problem by synthesizing all available studies.

GENDER DIFFERENCES IN MATHEMATICS PERFORMANCE

A major meta-analysis of studies of gender differences in mathematics performance surveyed 100 studies, representing the testing of more than 3 million people (Hyde et al., 1990). Averaged over all samples of the general population, $d = -0.05$, a negligible difference favoring females.

An independent meta-analysis confirmed the results of the first meta-analysis (Hedges & Nowell, 1995). It found effect sizes for gender differences in mathematics performance ranging between 0.03 and 0.26 across large samples of adolescents, all differences in the negligible to small range. Results from the International Assessment of Educational Progress (IAEP) also confirm that gender differences in mathematics performance are small across numerous countries, including Hungary, Ireland, Israel, and Spain (Beller & Gafni, 1996).

For issues of the underrepresentation of women in the physical sciences, however, this broad assessment of the magnitude of gender differences is probably less useful than an analysis by both age and cognitive level tapped by the mathematics test. These results from one meta-analysis are shown in Table 10.1. Ages were grouped roughly into elementary school (ages 5–10 years), middle school (11–14), high school (15–18), and college age (19–25). Insufficient studies were available for older ages to compute mean effect sizes.

Cognitive level of the test was coded as assessing either simple computation (requires the use of only memorized math facts, e.g., $7 \times 8 = 56$), conceptual (involves analysis or comprehension of mathematical ideas), problem solving (involves extending knowledge or applying it to new situations), or mixed. The results indicated that girls outperform boys by a small margin in computation in elementary school and middle school, and there is no gender difference in high school. For understanding of mathematical concepts, there is no gender difference at any age level. For problem solving, there is no gender difference in elementary or middle school but a small gender difference favoring males emerges in high school and the college years. There are no gender differences, then, or girls perform better, in all areas except problem solving beginning in the high school years.

This gender difference in problem solving favoring males deserves attention because problem solving is essential to success in occupations in the physical sciences. Perhaps the best explanation for this gender difference, in view of the absence of a gender difference at earlier ages, is that it is a result of gender differences in course choice, that is, the tendency of girls not to select optional advanced mathematics courses and science courses in high school. The failure to take advanced science courses may be particularly crucial because nonreform mathematics curricula often do not teach problem solving, whereas it typically is taught in chemistry and physics.

GENDER DIFFERENCES IN SPATIAL ABILITY

Spatial ability tests may tap any of several distinct skills: spatial visualization (finding a figure in a more complex one, like hidden-figures tests); spatial perception (identifying the true vertical or true horizontal when there is distracting information, e.g., the rod-and-frame task); and mental rotation (mentally rotating an object in three dimensions). Two meta-analyses are available on the question of gender differences in spatial performance. One found that the magnitude of gender differences varied substantially across the different types of spatial performance: $d = 0.13$ for spatial visualization, 0.44 for spatial perception, and 0.73 for mental rotation, all effects favoring males (Linn & Petersen, 1985). The last difference is large and potentially influential. The other meta-analysis found $d = 0.56$ for mental rotation (Voyer, Voyer, & Bryden, 1995), a somewhat smaller effect but nonetheless a substantial one. Gender differences in spatial performance—specifically, mental rotation—are important because mental rotation is crucial to success in fields such as engineering, chemistry, and physics.

GENDER DIFFERENCES IN VERBAL ABILITY

A meta-analysis of studies of gender differences in verbal ability indicated that, overall, the difference was so small as to be negligible, $d = -0.11$

(Hyde & Linn, 1988). The negative value indicates better performance by females, but the magnitude of the difference is quite small. There are many aspects to verbal ability, of course. When analyzed according to type of verbal ability, the results were as follows: for vocabulary, $d = -0.02$; for reading comprehension, $d = -0.03$; for speech production, $d = -0.33$; and for essay writing, $d = -0.09$. The gender difference in speech production favoring females is the largest and confirms females' better performance on measures of verbal fluency (not to be confused with measures of talking time). The remaining effects range from small to zero. Moreover, the magnitude of the effect was consistently small at all ages. Overall, then, gender differences in verbal ability are tiny and, if anything, favor females on measures such as essay writing and speech production, which should contribute to success in science. A second meta-analysis confirmed these findings using somewhat different methods (Hedges & Nowell, 1995).

In summary, then, when we look at three abilities that are essential for success in the physical sciences—mathematical ability, spatial ability, and verbal ability—the data provide no support for the notion that women lack the necessary abilities. Overall, girls and women score nearly the same as boys and men on mathematical tests, although a gender difference in complex problem solving emerges beginning in high school and deserves attention. Males score higher on tests of spatial ability—specifically, three-dimensional mental rotation—but, as I explain in a later section, it is quite possible to train and improve spatial ability. Finally, if anything, women have a slight advantage in verbal ability.

In the next section I review evidence on sociocultural forces that seem to contribute to gender differences in abilities. In particular, I examine the gender difference in mathematical problem solving beginning in high school and the gender difference in spatial ability.

SOCIOCULTURAL INFLUENCES ON GENDER DIFFERENCES IN MATHEMATICAL AND SPATIAL ABILITIES

The evidence on social and cultural influences on gender differences in mathematical and spatial abilities is plentiful and varied. I consider five categories of evidence: research on family, neighborhood, peer, and school influences; stereotype threat research; training studies; cross-cultural analyses; and trends over time.

Family, Neighborhood, Peer, and School Influences

Abundant evidence exists for the multiple influences of parents, peers, and the schools on children's development. Here I focus on these influences specifically in the domains of abilities and academic performance. A limitation to some of these studies is that they report simply a correlation, for example, between parents' estimates of the child's mathematics ability and the child's score on a standardized test. From this correlation, we cannot infer the direction of causality with complete certainty. We cannot tell whether the parents' beliefs in the child influence the child's performance or whether the opposite process occurs—that children's test scores influence their parents' estimates of abilities. Moreover, it may be that both processes occur in reciprocal fashion.

Numerous studies have confirmed the finding that parents' expectations for their children's academic abilities and success predict the children's self-concept of their own ability and their subsequent performance (e.g., Bleeker & Jacobs, 2004; Eccles, 1994). When engaged in a science task—playing with magnets—mothers talk about the science process (e.g., use explanations, generate hypotheses) more with boys than with girls (Tenenbaum, Snow, Roach, & Kurland, 2005). Moreover, the amount of mothers' science-process talk predicts children's comprehension of readings about science 2 years later. Observations of parents and children using interactive science exhibits at a museum showed that parents were three times more likely to explain science to boys than to girls (Crowley, Callanan, Tenenbaum, & Allen, 2001).

Sociologists and psychologists have also studied neighborhood effects, that is, effects outside the home that have an impact on children, such as living in a high-risk neighborhood. One well-sampled study of children in kindergarten and first grade found that boys' gains in math reasoning were more sensitive, both positively and negatively, to neighborhood resources than were girls' (Entwisle, Alexander, & Olson, 1994). By middle school, when children were tracked for math classes, boys in the high track were outscoring girls in that track, even though gender differences were not significant in the full sample. The researchers traced this greater variability for boys to their early sensitivity to neighborhood effects. Stronger neighborhood effects may occur for boys because boys spend more time outside the home than girls do. The researchers found that in the summer after first grade, boys were more likely, compared with girls, to be monitored less closely by their parents, to go to recreation centers, and to play organized sports, whereas girls were more likely to play in the house. Similar patterns continued in middle school. These differential experiences may give boys more spatial experience and, in complex games, more spatial and mathematical experience. Consistent with this idea, one study found that by age 11 boys show greater spatial knowledge than girls, as demonstrated, for example, by making a map of a familiar area. The researcher attributed this difference to girls' lesser experience of roaming over their environment, which in turn, he believes, is the result of parents' greater restrictions placed on girls (Matthews, 1986, 1987).

Children and adolescents are strongly influenced by peers. Here I focus on peer influence on academic performance and motivation. Children do, indeed, stereotype mathematics as masculine. In one study, girls rated adult men as liking and as being better at math than women, although they rated boys and girls as equal on these variables (J. Steele, 2003). By middle adolescence, girls generally receive less peer support for science activities than boys do (Stake & Nickens, 2005). Science enrichment programs can be helpful in counteracting these effects, by giving girls a science-supportive peer network (Stake & Nickens, 2005).

Schools may exert their influence in multiple ways, including teachers' attitudes and behaviors, curriculum, ability grouping, and sex composition of the classroom. The availability of hands-on laboratory experiences is especially critical for learning in the physical sciences in middle school and high school. An important point is that, although laboratory experiences do not improve the physical science achievement of boys, they do improve the achievement of girls, thereby helping to close the gender gap in achievement in the physical sciences (Burkam, Lee, & Smerdon, 1997; Lee & Burkam, 1996). In science and mathematics classes, teachers are more likely to encourage boys than girls to ask questions and to explain (American Association of University Women, 1995; Jones & Wheatley, 1990; Kelly, 1988). In one study of high school geometry classrooms, teachers directed 61% of their praise comments to boys and 55% of their high-level open questions to boys (Becker, 1981). Experiences such as these are thought to give children a deeper conceptual knowledge of and more interest in science. Moreover, even an apparently small gender difference of teachers spending 44% of their time with girls and 56% with boys works out to 1,800 more hours with boys over a child's school career of 15,000 hours (Kelly, 1988).

Not only do school environments differ for boys and girls, but home environments do as well. Presence of a computer in the home for educational purposes and home computer use predict performance on standardized tests of mathematics (Downey & Yuan, 2005). Compared with high school girls, high school boys are more likely to have a computer in their home for educational purposes and are more likely to use it.

Of course, schools do not exert absolute power over completely passive students. Students exercise choice in school activities. Crucial to this discussion is their choice in high school to take advanced mathematics and science courses. The gender gap in mathematics course enrollment has narrowed over the past decade, so that by 1998 girls were as likely as boys to have taken advanced mathematics courses, including Advanced Placement or International Baccalaureate (AP or IB) calculus (National Science Foundation, 2005). Girls were actually slightly more likely than boys to take advanced biology (40.8% of girls, 33.8% of boys), AP biology (5.8% of girls, 5.0% of boys), and chemistry (59.2%, 53.3%). Boys, however, were more likely to take AP chemistry (3.3% of boys, 2.6% of girls) and physics (31% of boys, 26.6% of girls) and were twice as likely to take AP physics (2.3% of boys, 1.2% of girls; National Science Foundation, 2005). The science pipeline heading toward physics, then, begins to leak early as fewer girls take the necessary high school courses to prepare themselves for college-level physics. It is beyond the scope of this chapter to review what psychologists know about the reasons why children choose or do not choose to take challenging math and science courses. Readers wanting more information can look to a massive program of research conducted by Jacquelynne Eccles (e.g., Eccles, 1994).

Stereotype Threat

Stereotype threat is an influence that may occur in an actual testing situation. It was initially identified and theorized to explain differences in test performance between talented Black and White college students (C. M. Steele, 1997). Psychologist Claude Steele (1997) believes that a negative stereotype about one's group leads to self-doubt and other processes, which then damage academic performance. This concept was quickly extended from stereotypes about Blacks' intellectual inferiority to stereotypes about women's deficiencies in mathematics (Quinn & Spencer, 2001). In one experiment, male and female college students with equivalent math backgrounds were tested (Spencer, Steele, & Quinn, 1999). Half were told that the math test had shown gender differences in the past, and half were told that the test had been shown to be gender fair—that men and women had performed equally on it. Among those who were led to believe that the test was gender fair, there were no gender differences in performance, but among those who believed it showed gender differences, women underperformed compared with men.

No one has yet conducted a meta-analysis of these stereotype-threat studies, so the size of the effect is unknown. Some studies show large effects ($d = 1.35$, Johns, Schmader, & Martens, 2005; $d = 0.67$, Quinn & Spencer, 2001).

Additional research has elaborated the findings on gender and stereotype threat. For example, under stereotype-threat conditions when solving difficult math problems, women's ability to formulate problem-solving strategies is reduced, compared with low-threat conditions (Quinn & Spencer, 2001). On a more positive note, the presence of a math-competent female role model eliminates the stereotype-threat effect on women's math performance (Marx & Roman, 2002).

The stereotype-threat research carries two implications here. First, if a simple manipulation of instructions can produce or eliminate gender differences in performance on a mathematics exam, the notion of fixed gender differences in math ability is called into serious question. Second, stereotype threat is a result of cultural factors—specifically, gender stereotypes about female inferiority at mathematics—and thus provides evidence of sociocultural influence on gender differences in mathematics performance.

Training Studies

Environmental input is essential to the development of spatial and mathematical abilities (Baenninger & Newcombe, 1995; Newcombe, 2002; Spelke, 2005). Babies are not born knowing how to work calculus problems. Children acquire these skills through schooling and other experiences.

A meta-analysis found that spatial ability can indeed be improved with training, with effect sizes ranging between $d = 0.40$ to 0.80, depending on the length and specificity of the training (Baenninger & Newcombe, 1989). The effects of training were similar for males and females; that is, both groups benefited about equally from the training, and there was little evidence that the gender gap was closed or widened by training. A more recent study showed that the gender difference could be eliminated by carefully conceptualized training (Vasta, Knott, & Gaze, 1996). Unfortunately, most school curricula contain little or no emphasis on spatial learning. Girls, especially, could benefit greatly from such a curriculum.

Cross-Cultural Analyses

The IAEP tested the math and science performance of 9- and 13-year-olds in 20 countries around the world. The effect sizes for gender differences for selected countries are shown in Table 10.2 (Beller & Gafni, 1996). Focusing first on the results for mathematics, we see that the gender differences are small in all cases. Most important, effect sizes are positive (favoring males) in some countries, negative (favoring females) in other countries, and essentially zero in several others. The Trends in International Mathematics and Science Study (TIMSS, 2003; formerly the Third International Mathematics Study) found similar results, with some positive and some negative effect sizes, and most < .10. In the TIMSS data for eighth graders, the magnitude of the gender difference was 0.09 in Chile (country average score = 379), 0.02 in the United States (country average score = 502), 0.01 in Japan (country average score = 569), and –0.05 in Singapore (country average score = 611). That not only the magnitude but also the direction of gender differences in mathematics performance varies from country to country is powerful testimony to the importance of sociocultural factors in shaping those differences. Perhaps most important, though, the gender difference is very small in most nations.

Focusing next on the results for science performance (Table 10.2), we can see that the effect sizes more consistently favor males and are somewhat larger, although not large for any nation. When the results are broken down by science, gender differences are smaller in life sciences knowledge (0.11

TABLE 10.2
Effect Sizes for Gender Differences in Mathematics and Science Test Performance Across Countries

Country	Mathematics		Science	
	9 years	13 years	9 years	13 years
Hungary	−0.03	−0.02	0.09	0.25
Ireland	−0.06	0.19	0.20	−0.31
Israel	0.16	0.15	0.23	0.24
Korea	0.28	0.10	0.39	0.31
Scotland	−0.01	−0.02	−0.01	0.20
Spain	0.01	0.18	0.25	0.24
Taiwan	0.03	0.02	0.25	0.08
United States	0.05	0.04	0.09	0.29
All countries	0.04	0.12	0.16	0.26

Note. From "The 1991 International Assessment of Educational Progress in Mathematics and Sciences: The Gender Differences Perspective," by M. Beller and N. Gafni, 1996, *Journal of Educational Psychology, 88,* Table 2 (p. 370) and Appendix (p. 377). Copyright 1996 by the American Psychological Association.

and 0.20 at ages 9 and 13, respectively, averaged over all countries) and somewhat larger for physical sciences (0.22 and 0.33; Beller & Gafni, 1996).

It is important to note that cross-cultural differences in mathematics performance are enormous compared with gender differences in any one country. For example, in one cross-national study of fifth graders, American boys (M = 13.1) performed better than American girls (M = 12.4) on word problems, but fifth-grade Taiwanese girls (M = 16.1) and Japanese girls (M = 18.1) performed far better than American boys (Lummis & Stevenson, 1990). Culture is considerably more important than gender in determining mathematics performance.

In perhaps the most sophisticated analysis of cross-national patterns of gender differences in mathematics performance, the researchers found that, across nations, the magnitude of the gender difference in mathematics performance for eighth graders correlated significantly with a variety of measures of gender stratification in the countries (Baker & Jones, 1993). For example, the magnitude of the gender difference in math performance correlated −.55, across nations, with the percentage of women in the workforce in those nations. That is, the more that women participate in the labor force (an index of gender equality), the smaller the gender difference in mathematics achievement.

Trends Over Time

Changes over time in patterns of gender differences in mathematics and science aptitude and achievement provide further evidence of sociocultural influence. If the gender differences were completely determined by biological factors, they would remain static over time.

TABLE 10.3
Degrees in Science and Engineering Awarded to Women, 1966 and 2000

Degree	1966 (%)	2000 (%)
Bachelor's degree		
Engineering	0.4	20.5
Physical sciences	14.0	41.1
Earth, atmospheric, and oceanic sciences	9.4	40.0
Mathematical and computer sciences	33.2	32.7
Biological sciences	25.0	55.8
Doctoral degree		
Engineering	0.3	15.8
Physical sciences	4.5	24.6
Earth, atmospheric, and ocean sciences	3.0	30.6
Mathematical and computer sciences	6.1	21.0
Biological sciences	12.0	42.6

Note. Data from National Science Foundation (2002).

Evidence of trends over time comes from gendered patterns of degrees earned in science and engineering (see Table 10.3). For example, in 1966 women earned only 4.5% of the doctoral degrees in the physical sciences. By 2000, this percentage had risen to 24.6%. As societal opportunities have increased, women's achievements in these areas have increased. Moreover, if women do not have the "right stuff" to succeed in the sciences, it is difficult to understand how they can be earning 24.6% of the doctorates in the physical sciences. This substantial presence of women among doctoral recipients in the physical sciences also calls into question the import of greater male variability as an explanation for the gender gap in science occupations. It is clear that there are substantial numbers of women at the upper tail of the ability distribution, as evidenced by the fact that they are able to earn PhDs in the physical sciences.

IMPLICATIONS: HOW CAN WE CLOSE THE GENDER GAP IN THE PHYSICAL SCIENCES?

One conclusion of this review is that, overall, there are no gender differences in math performance, but a gender difference favoring males in complex problem solving does emerge in high school. Mathematical problem solving is crucial to success in the physical sciences, so this gap must be addressed. The evidence also indicates a gender gap in favor of males in spatial ability, specifically in mental rotation. This ability, too, is crucial to success in the physical sciences and must be addressed.

A major part of the gender gap in mathematical problem solving is due to differential patterns of course choice for boys and girls. Until very recently, girls were less likely than boys to take advanced math courses in high school, and they still are less likely to take physics. Students learn complex problem solving both in math courses and in science courses, so if girls take fewer courses in either area, they will not score as high. Therefore, one important recommendation is to require all college-bound students to take 4 years of high school math and 4 years of high school science. It is a win–win solution! Both high schools and colleges can contribute to this effort. Colleges and universities, for example, can require these courses for admission. This will help to ensure that young women do not enter college with a science deficit.

In addressing the gender gap in spatial ability, the two most important points to recognize are, first, that spatial ability can be trained and learned and, second, that with only a few exceptions, there is no spatial curriculum in the schools. An important recommendation, therefore, is to institute a spatial curriculum in the schools, beginning in the elementary grades. There is reason for optimism about the potential for success. The most recent development is multimedia software that provides training in three-dimensional spatial visualization skills (Gerson, Sorby, Wysocki, & Baartmans, 2001). It has been used successfully with first-year engineering students. It is important to note that there were improvements in the retention of women engineering students who took the spatial visualization course; without the course, the retention rate for women was 47% whereas with the course it was 77%.

SUMMARY

Sociocultural forces play a major role in shaping gender differences in math and science abilities and occupational success in science careers. Parents are part of the puzzle insofar as they encourage sex-typed activities for their children and math and science are seen as more appropriate for boys than girls. Neighborhood effects may be another piece to the puzzle, although there is less research here. Peers are influential, and children tend to believe that men are better than women at math. The schools certainly influence students' learning and performance; research has documented systematic, subtle differences in the ways that teachers treat boys compared with girls in science and mathematics classrooms.

Cross-cultural research demonstrates that not only the magnitude but also the direction of gender differences in mathematics performance varies across nations. In no country is the gender difference large. Moreover, the magnitude of the gender difference correlates with measures of gender equality in the country.

Finally, I would be remiss not to mention the impact of sex discrimination. A substantial body of research has investigated sex discrimination in the evaluation of work and experiences of sex discrimination among women in science and engineering fields (Swim, Borgida, Maruyama, & Myers, 1989). Women in science do report significant sex discrimination, and these experiences likely shape the direction their careers take (J. Steele, James, & Barnett, 2002).

Certainly there are caveats to these conclusions because of methodological limitations of individual studies. Nonetheless, most of the conclusions are based on very sound science, and some are based on meta-analyses of large bodies of research. Of the factors listed, no single one by itself has been shown to be the determining factor in women's underrepresentation in careers in the physical sciences. Each may have a small effect, and these small effects add up. Moreover, some factors may be influential in some women's lives and other factors are the key for other women.

REFERENCES

American Association of University Women. (1995). *How schools shortchange girls.* Washington, DC: Author.

Baenninger, M., & Newcombe, N. (1989). The role of experience in spatial test performance: A meta-analysis. *Sex Roles, 20,* 327–344.

Baenninger, M., & Newcombe, N. (1995). Environmental input to the development of sex-related differences in spatial and mathematical ability. *Learning and Individual Differences, 7,* 363–379.

Baker, D. P., & Jones, D. P. (1993). Creating gender equality: Cross-national gender stratification and mathematical performance. *Sociology of Education, 66,* 91–103.

Becker, J. R. (1981). Differential treatment of females and males in mathematics classes. *Journal for Research in Mathematics Education, 12,* 40–53.

Beller, M., & Gafni, N. (1996). The 1991 International Assessment of Educational Progress in mathematics and sciences: The gender differences perspective. *Journal of Educational Psychology, 88,* 365–377.

Bleeker, M. M., & Jacobs, J. E. (2004). Achievement in math and science: Do mothers' beliefs matter 12 years later? *Journal of Educational Psychology, 96,* 97–109.

Burkam, D. T., Lee, V. E., & Smerdon, B. A. (1997). Gender and science learning early in high school: Subject matter and laboratory experiences. *American Educational Research Journal, 34,* 297–331.

Cohen, J. (1988). *Statistical power analysis for the behavioral sciences* (2nd ed.). Hillsdale, NJ: Erlbaum.

Crowley, K., Callanan, M. A., Tenenbaum, H. R., & Allen, E. (2001). Parents explain more often to boys than to girls during shared scientific thinking. *Psychological Science, 12,* 258–261.

Downey, D. B., & Yuan, A. (2005). Sex differences in school performance during high school: Puzzling patterns and possible explanations. *Sociological Quarterly, 46,* 299–321.

Eccles, J. S. (1994). Understanding women's educational and occupational choices: Applying the Eccles et al. model of achievement-related choices. *Psychology of Women Quarterly, 18,* 585–610.

Entwisle, D. R., Alexander, K. L., & Olson, L. S. (1994). The gender gap in math: Its possible origins in neighborhood effects. *American Sociological Review, 59,* 822–838.

Gerson, H., Sorby, S. A., Wysocki, A., & Baartmans, B. J. (2001). The development and assessment of multimedia software for improving 3-D spatial visualization skills. *Computer Applications in Engineering Education, 9,* 105–113.

Hedges, L. V., & Becker, B. J. (1986). Statistical methods in the meta-analysis of research on gender differences. In J. S. Hyde & M. C. Linn (Eds.), *The psychology of gender: Advances through meta-analysis* (pp. 14–50). Baltimore: Johns Hopkins University Press.

Hedges, L. V., & Nowell, A. (1995, July 7). Sex differences in mental test scores, variability, and numbers of high-scoring individuals. *Science, 269,* 41–45.

Hyde, J. S., Fennema, E., & Lamon, S. J. (1990). Gender differences in mathematics performance: A meta-analysis. *Psychological Bulletin, 107,* 139–155.

Hyde, J. S., & Linn, M. C. (1988). Gender differences in verbal ability: A meta-analysis. *Psychological Bulletin, 104,* 53–69.

Johns, M., Schmader, T., & Martens, A. (2005). Knowing is half the battle: Teaching stereotype threat as a means of improving women's math performance. *Psychological Science, 16,* 175–179.

Jones, M. G., & Wheatley, J. (1990). Gender differences in teacher–student interactions in science classrooms. *Journal of Research in Science Teaching, 27,* 861–874.

Kelly, A. (1988). Gender differences in teacher–pupil interactions: A meta-analytic review. *Research in Education, 39,* 1–23.

Lee, V. E., & Burkam, D. T. (1996). Gender differences in middle grade science achievement: Subject domain, ability level, and course emphasis. *Science Education, 80,* 613–650.

Linn, M. C., & Petersen, A. C. (1985). Emergence and characterization of sex differences in spatial ability: A meta-analysis. *Child Development, 56,* 1479–1498.

Lummis, M., & Stevenson, H. W. (1990). Gender differences in beliefs and achievement: A cross-cultural study. *Developmental Psychology, 26,* 254–263.

Marx, D. M., & Roman, J. S. (2002). Female role models: Protecting women's math test performance. *Personality and Social Psychology Bulletin, 28,* 1183–1193.

Matthews, M. H. (1986). The influence of gender on the environmental cognition of young boys and girls. *Journal of Genetic Psychology, 147,* 295–302.

Matthews, M. H. (1987). Sex differences in spatial competence: The ability of young children to map "primed" unfamiliar environments. *Educational Psychology, 7,* 77–90.

National Science Foundation. (2002). *Science and engineering degrees, 1966–2000* (NSF 02-327). Retrieved July 5, 2005, from http://www.nsf.gov/sbe/srs/stats.htm

National Science Foundation. (2005). *Science and engineering indicators 2004: Elementary and secondary education, mathematics and science coursework and student achievement.* Retrieved July 5, 2005, from http://www.nsf.gov/statistics/seind04/

Newcombe, N. S. (2002). The nativist–empiricist controversy in the context of recent research on spatial and quantitative development. *Psychological Science, 13*, 395–401.

Quinn, D. M., & Spencer, S. J. (2001). The interference of stereotype threat with women's generation of mathematical problem-solving strategies. *Journal of Social Issues, 57*, 55–72.

Spelke, E. S. (2005). Sex differences in intrinsic aptitude for mathematics and science? A critical review. *American Psychologist, 60*, 950–958.

Spencer, S. J., Steele, C. M., & Quinn, D. M. (1999). Stereotype threat and women's math performance. *Journal of Experimental Social Psychology, 35*, 4–28.

Stake, J. E., & Nickens, S. D. (2005). Adolescent girls' and boys' science peer relationships and perceptions of the possible self as scientist. *Sex Roles, 52*, 1–12.

Steele, C. M. (1997). A threat in the air: How stereotypes shape intellectual identity and performance. *American Psychologist, 52*, 613–629.

Steele, J. (2003). Children's gender stereotypes about math: The role of stereotype stratification. *Journal of Applied Social Psychology, 33*, 2587–2606.

Steele, J., James, J. B., & Barnett, R. C. (2002). Learning in a man's world: Examining the perceptions of undergraduate women in male-dominated academic areas. *Psychology of Women Quarterly, 26*, 46–50.

Swim, J., Borgida, E., Maruyama, G., & Myers, D. G. (1989). Joan McKay versus John McKay: Do gender stereotypes bias evaluations? *Psychological Bulletin, 105*, 409–429.

Tenenbaum, H. R., Snow, C. E., Roach, K. A., & Kurland, B. (2005). Talking and reading science: Longitudinal data on sex differences in mother–child conversations in low-income families. *Applied Developmental Psychology, 26*, 1–19.

Trends in International Mathematics and Science Study. (2003). *Highlights from the Trends in International Mathematics and Science Study (TIMSS) 2003.* Retrieved July 13, 2005, from http://nces.ed.gov/pubs2005/timss03

Vasta, R., Knott, J. A., & Gaze, C. E. (1996). Can spatial training erase the gender differences on the water-level task? *Psychology of Women Quarterly, 20*, 549–568.

Voyer, D., Voyer, S., & Bryden, M. P. (1995). Magnitude of sex differences in spatial abilities: A meta-analysis and consideration of critical variables. *Psychological Bulletin, 117*, 250–270.

11

THE SEEDS OF CAREER CHOICES: PRENATAL SEX HORMONE EFFECTS ON PSYCHOLOGICAL SEX DIFFERENCES

SHERI A. BERENBAUM AND SUSAN RESNICK

There is not a simple answer to the question "Why aren't more women in science?" Much of the debate surrounding this question has focused on whether sex differences in cognition are responsible for the underrepresentation of women in science. This focus is unsatisfying because no psychological characteristic is determined by a single factor, and because it ignores sex differences in other careers. Therefore, in this chapter we focus on the question "How are sex differences in cognition and other characteristics related to the differential representation of women and men in scientific and social service careers?"

We focus on individual characteristics contributing to career choices and outcomes, recognizing that individuals operate in a social environment and that individual behavior is affected by social context. Thus, sex differ-

Some of the work described in this chapter was supported by Grant HD19644 from the National Institutes of Health.

ences in career choices and outcomes are not completely attributable to individuals but also reflect contributions from social institutions, including sex discrimination and issues related to juggling work and family, such as access to child care.

Within the realm of individual characteristics, we focus on how biological factors may affect individual differences in career choices, directly (e.g., through interests or abilities) and indirectly (e.g., through modification of the social environment). Social factors (e.g., parent attitudes and media portrayals) may affect individual characteristics related to career choice, but individuals are not passive recipients of the social environment. We propose that sex-related career choices and outcomes arise through the mediating and moderating effects of socialization on sex-hormone-influenced individual differences in behavioral development. For example, prenatal sex hormone exposure may influence specific cognitive abilities and interests later in life, but social factors moderate the expression of these individual differences in abilities and interests.

HOW MIGHT SEX HORMONES MEDIATE SEX DIFFERENCES IN CAREER CHOICE?

Studies in nonhuman mammals show that sex hormones present early in life affect the development of juvenile and adult sex-typical physical and behavioral characteristics. In particular, exposure to high levels of androgens (characteristic of males) increases behavior that is typical of males and inhibits behavior that is typical of females (Becker, Breedlove, Crews, & McCarthy, 2002; Wallen, 2005). For example, female monkeys treated prenatally with androgen are masculinized in sexual behavior, rough play, grooming, and some learning abilities.

A key question has been whether similar effects occur in people. It is, of course, unethical to manipulate hormones in human beings, but it is possible to study people whose hormones are atypical for their sex. The group most frequently studied is females with congenital adrenal hyperplasia (CAH), who have a genetic defect in an enzyme controlling cortisol production, leading them to produce high levels of androgens starting early in gestation. Although these girls are usually born with some masculinization of their external genitals, they are reared as girls, treated with cortisol to reduce their androgen levels, and have surgery to feminize their external genitals. If prenatal androgens contribute to human psychological sex differences, then females with CAH should tend to be psychologically more masculine and less feminine than a comparison group of females without CAH. And some studies suggest that females with CAH, on average, may have masculinized interests and specific cognitive abilities, although they have female-typical gender identity and think of themselves as female (for reviews, see Berenbaum,

2001; Cohen-Bendahan, van de Beek, & Berenbaum, 2005; Meyer-Bahlburg, 2001).

People often misunderstand what it means for biology to affect behavior, so we emphasize three points as we consider how prenatal hormones influence human psychological characteristics related to career choices. First, hormones are unlikely to have absolute and fixed effects on any characteristic, physical or psychological. The biological effects of hormones depend on genes, and gene expression depends on the physical and social environment. Second, hormone effects do not have to be direct. Hormones may influence specific brain structures that underlie specific behaviors, but it is unlikely that characteristics related to career choice are simply subserved by a single brain area that is fixed at birth. Third, political equality does not depend on biological equality. All people should be treated the same under the law without regard to the origins of the differences among them.

ANDROGEN EFFECTS ON PREDICTORS OF CAREER CHOICE

Females with CAH are of considerable value in understanding biological contributors to sex differences in career choices, because their female sex of rearing combined with atypical androgen levels produce a separation of social and hormonal influences on these choices. The key question is whether females with CAH are more likely than those without CAH to pursue science careers, but there are no data about career choices of women with CAH. Nevertheless, there is good evidence that androgens influence characteristics related to career outcomes, including specific aspects of cognitive abilities, interests, personality, and social interactions.

Androgens and Cognition

There is continued controversy about the nature and size of cognitive sex differences, but even "gender similarities" theorists acknowledge some average differences between males and females, particularly with respect to spatial abilities. What is the origin of these differences? Socialization contributes to average sex differences in some cognitive abilities, for example, through stereotypes emphasizing women's inferiority in spatial and math abilities and promotion of spatial experiences in boys but not girls. There is also good evidence that early androgens facilitate the development of spatial ability.

Several studies indicated that females with CAH perform better on some (but not all) spatial tests than their unaffected sisters in childhood, adolescence, and adulthood (Berenbaum, 2001; Hampson, Rovet, & Altmann, 1998; Hines et al., 2003; Resnick, Berenbaum, Gottesman, & Bouchard, 1986). But, there are some inconsistencies in results, perhaps because of methodological limitations, particularly small numbers of participants (50 partici-pants per group are needed to ensure that differences in the population are detected in any specific sample, a difficult goal in studies of these clinical samples). Data from another clinical condition are consistent with findings in females with CAH: Males with low early androgen levels (due to idio-pathic hypogonadotropic hypogonadism) have lower spatial ability than con-trol males.

Thus, prenatal androgen exposure likely contributes to sex differences in spatial ability. Because spatial ability may be important for problem solv-ing in science and engineering (e.g., visualization of three-dimensional shape problems), it will be interesting to see whether enhanced spatial abilities of women with CAH increase the likelihood that they choose careers in sci-ence and engineering. Although good spatial and mathematical skills are important for some scientific careers, we propose that other sex-related characteristics may be more important than cognition in career choice and success.

Androgens and Activity Interests

Another finding from studies of females with CAH is that prenatal androgens are strongly associated with later sex-related interests, including childhood toy preferences, adolescent activity interests, and adult hobbies (Berenbaum, 2004; Cohen-Bendahan et al., 2005). In childhood, girls with CAH play more with boys' toys and less with girls' toys than do girls without CAH. This is seen in many ways, including direct observations and reports. For example, girls with CAH ages 3 to 12 years played with boys' toys 1.5 to 2 times as much as comparison girls. When choosing a toy to keep across several occasions, 43% of girls with CAH chose a car or truck at any time, whereas 4% of control girls (and 76% of boys) did. Androgens continue to exert large effects on gender-typical activities beyond childhood. As mea-sured by self-reports and parent reports, adolescent and adult females with CAH are very interested in male-typical activities, such as electronics and sports, and not as interested in female-typical activities, such as fashion and crafts.

The sex-atypical interests of females with CAH extend to vocational interests. For example, teenage girls with CAH express interest in male-typical careers, such as engineer, architect, and airline pilot, whereas their unaffected sisters express interest in female-typical careers, such as X-ray tech-nician, ice skater, and hairstylist. Vocational interests are stable across time in typical males and females (Low, Yoon, Roberts, & Rounds, 2005), so it seems likely that adult females with CAH will maintain interest in male-typical occupations, and we are studying this.

Androgen effects on interests are very large, and there is little overlap in measured interests of females with CAH and typical females. However,

CAH is not a perfect experiment, and differences between females with and without CAH might be due to factors besides androgen. The most popular alternative explanation is that sex-atypical behavior is caused by parents treating girls with CAH like boys because of their masculinized external genitals. But, that does not appear to occur. Girls with CAH do not play more with boys' toys when their parents are present (Nordenström, Servin, Bohlin, Larsson, & Wedell, 2002), and they are encouraged by their parents to play with girls' rather than boys' toys (Pasterski et al., 2005). Neither have other alternative explanations, including other hormonal changes and effects of chronic illness, been supported by data (Berenbaum, 2004).

Androgens and Sex-Typed Personal Characteristics and Social Interactions

Androgens might influence career choices through other characteristics, too. Females with CAH differ from those without CAH on a number of sex-related behaviors that might be important for career outcomes (Berenbaum & Resnick, 1997; Leveroni & Berenbaum, 1998; Meyer-Bahlburg, 1999). They are less interested in babies and more likely to report that they would use physical aggression in hypothetical conflict situations. Other aspects of personality and social behavior also show sex differences, including emotional expressivity, intimacy in friendships, concern with physical appearance, and thrill-seeking (Ruble, Martin, & Berenbaum, 2006; Wood & Eagly, 2002), but we do not yet know whether these dimensions are influenced by prenatal androgens and whether these traits are related to career choices.

Androgens affect not only individual behavior but also sex-related social interactions, including childhood peer preferences and adult sexual orientation. Girls with CAH are more likely than girls without CAH to report that they play with and prefer boys as playmates (Hines & Kaufman, 1994; Servin, Nordenström, Larsson, & Bohlin, 2003). As adults, they are less likely to be sexually attracted to men and more likely to be sexually attracted to women than are their sisters without CAH (Hines, Brook, & Conway, 2004; Zucker et al., 1996). Because peers play a large role in gender socialization (Maccoby, 1998; Ruble et al., 2006), variations in peer interactions could have long-lasting psychological effects, including impact on career choices.

SPECULATIONS ON THE PATHS FROM HORMONES TO CAREER CHOICE

Prenatal androgens thus affect some traits related to career outcomes, including aspects of spatial ability, activity interests, personal and social behavior, and social relationships—and part of the sex difference in career choices themselves could be explained by effects of prenatal androgens on these traits, although this awaits direct test. This answers part of the question we posed at the beginning of our essay: "How are sex differences in cognition and other characteristics related to the differential representation of women and men in scientific and social service careers?" But, it is important to remember that androgen effects on careers must be mediated through psychological (and perhaps physical) traits, including but not restricted to those we discussed. Although sex differences in brain structure and function are widespread even in humans (Resnick, 2006), a decision to become a scientist is not determined by androgens acting on specific brain regions during prenatal development. Instead, prenatal hormonal variations influence brain development and the way an individual approaches the postnatal environment, including the choice of career.

Nature via Nurture: Androgen Effects on Behavior Through the Environment

Findings from females with CAH are used to infer a role for biology in psychological sex differences, but they can do much more, helping to understand how nature and nurture work together, how a child's (biologically predisposed) preferences guide her down a pathway of social experiences that lead her to choose one career over another. Androgens affect psychological characteristics through a person's transactions with the (physical and social) environment. This is amply documented in other species, and even more likely to be the case in human beings for whom culture is crucial. In rodents, for example, hormones affect both the behavior of the organism and others' responses to it, including grooming by the mother and attraction of peers (Clark & Galef, 1998).

Such hormone–environment joint effects should—and can—be studied in people (although they have not yet). Consider a possible pathway by which prenatal androgens might affect adolescent career choice, illustrated in the hypothetical development of a girl with CAH. Her exposure to high levels of androgen leads her to prefer moving stimuli, and this is socially reinforced, with toy vehicles, action toys, and Legos provided by her parents partly in response to her request. Frequent and positive experiences with machines and construction toys lead to other behaviors as she matures: She moves around her environment, enjoys spatial activities and becomes good at them, recognizes her good spatial skills, sees herself as someone who can affect her environment, and seeks out school experiences consistent with her interests (e.g., she joins sports teams and computer club). These interests also produce social responses from others; for example, parents provide her with videogames, teachers encourage her to participate in the Geography Bee. If she spends more time with boys (as some girls with CAH appear to do), then her boy-typical interests will be reinforced, and she will be influenced more by the culture of boys' groups than that of girls' groups (e.g., boys' play involves physical contact and dominance hierarchies, whereas girls' groups emphasize cooperation and communication; Maccoby, 1998). Taken together, the child-selected and the adult- and peer-facilitated environment reinforce the girl's interests and abilities related to moving objects and spatial information, facilitating interests, abilities, and confidence related to science and math. A longitudinal study (with more clearly articulated hypotheses) could provide an empirical test of this speculative mechanism from androgen to career choice.

Implications for Increasing the Representation of Women in Science

Some people try to deny the existence of sex differences in cognition, interests, and personality, as well as possible biological contributions to individual differences in these behaviors, out of fear that they might be used to justify inequalities. It is often assumed that any biological contribution to women's underrepresentation in science would doom women to a continued disadvantage in these careers. But this reasoning is wrong! Furthering our understanding of biology should facilitate rather than hinder solutions to such inequities.

Biology is not destiny: For all characteristics studied, the differences between females with and without CAH are smaller than the sex differences, which means that androgens do not account for all of the average differences between males and females. This is consistent with other evidence showing the importance of social influences on psychological sex differences (Ruble et al., 2006; Wood & Eagly, 2002).

Moreover, even characteristics strongly shaped by genes through hormones (or other aspects of physiology) are typically modified by the environment. The potential for environmental modification of gene expression is the basis for much of social and public health policy and people's daily lives. Newborn screening for phenylketonuria is conducted so that genetically based intellectual disability can be prevented by dietary change. Successful treatments of disease depend on discovery of agents that modify how genes are expressed. Lifestyle changes, such as diet and exercise, reduce risk for cardiovascular disease and diabetes, and cosmetics and surgery modify physical attractiveness.

Environmental modification of gene expression extends to the brain and behavior and is especially important given the multifactorial nature of most behavioral traits. This is well documented in other species. For example, in monkeys, restricted experience with peers hinders sexual behavior (Wallen, 1996); in rats, mothers lick and groom male pups more than female pups, and these differences influence adult sexual behavior and stress responses through changes in gene expression (Moore, 1992). We know that people's experience affects their behavior: Psychoeducational interventions improve the abilities and social behaviors of children with Down syndrome and autism; psychological and pharmacologic therapies improve mental health. And the brain is changed by experiences, both naturally occurring ones and interventions. For example, aberrant emotional experience associated with maltreatment alters the ways in which the brain processes emotion (Pollak, Klorman, Thatcher, & Cicchetti, 2001); variations in taxi driving experience are associated with variations in the size of the hippocampus (Maguire et al., 2003); physical fitness reduces age-related structural and functional brain declines (Colcombe et al., 2004); and cognitive–behavioral therapy ameliorates anxiety disorders and depression by modulating brain activity (Goldapple et al., 2004; Paquette et al., 2003).

In general, factors producing behavioral differences are distinct from those that maintain or modify them. Social interventions might reduce or eliminate sex differences that are initially influenced by hormones. Conversely, social causes of sex differences do not imply easy interventions.

With respect to cognition, evidence shows that spatial ability can be improved with training and practice (Baenninger & Newcombe, 1989). Training of spatial skills is likely to have a bigger benefit than other interventions aimed at improving girls' spatial abilities such as giving them boys' toys. For example, a girl who starts out with low spatial ability because her prenatal androgens were low can increase her ability through experience with spatial materials. Training also improves spatial ability in boys and may not eliminate the sex difference, but our challenge is to provide all individuals with sufficient basic skills to excel in science and mathematics. Equipped with basic skill sets, all individuals will have a full range of career choices. Ability alone does not determine success in those chosen careers, because motivation, personality, and personal decisions regarding balance between family and work also contribute to an individual's advancement in her chosen profession.

We know little about changing other characteristics that are influenced by prenatal androgen and might relate to career choice, but it is reasonable to ask why we should change them. For example, what is to be gained by having girls play more with boys' toys? Girls already play with boys' toys, just not as much as boys do, so this would involve having them play almost exclusively with boys' toys and therefore not with girls' toys. In general, girls appear to have a wider range of interests than boys do, so this intervention would reduce girls' options, hardly a good solution.

It is also important to recognize that career outcome and its value depend not just on individual characteristics but on the actions of social institutions. Discrimination and unavailability of child care may reduce participation by women with the interest and talent for a scientific career. Women are underrepresented in fields that do not depend on sex-typed skills or interests, indicating that sex differences in career outcomes are influenced in considerable ways by these social forces. Debates focus on women's underrepresentation in science but generally exclude discussion of the underrepresentation of men in social service occupations, probably because of the greater status, prestige, and financial rewards of scientific than social service careers, and the greater power accorded to men than to women. Perhaps we could solve the problem of women's underrepresentation in science and math not just by increasing their representation in these professions but by valuing their contributions in other fields.

Returning to the question we posed at the beginning of our essay—"How are sex differences in cognition and other characteristics related to the differential representation of women and men in scientific and social service careers?"—we encourage expanding the discussion beyond the effect of cognitive sex differences on representation of men and women in science and engineering. It is important to consider factors leading to the greater representation of women in social service careers and the variety of noncognitive personal and social factors that contribute to the development, maintenance, and modification of career choices. Our challenge is to provide the educational training and social resources to allow both men and women access to the full range of career choices and development.

REFERENCES

Baenninger, M., & Newcombe, N. (1989). The role of experience in spatial test performance: A meta-analysis. *Sex Roles, 20,* 327–344.

Becker, J. B., Breedlove, S. M., Crews, D., & McCarthy, M. M. (Eds.). (2002). *Behavioral endocrinology* (2nd ed.). Cambridge, MA: MIT Press.

Berenbaum, S. A. (2001). Cognitive function in congenital adrenal hyperplasia. *Endocrinology and Metabolism Clinics of North America, 30,* 173–192.

Berenbaum, S. A. (2004). Androgen and behavior: Implications for the treatment of children with disorders of sexual differentiation. In O. H. Pescovitz & E. A. Eugster (Eds.), *Pediatric endocrinology: Mechanisms, manifestations, and management* (pp. 275–284). Philadelphia: Lippincott Williams & Wilkins.

Berenbaum, S. A., & Resnick, S. M. (1997). Early androgen effects on aggression in children and adults with congenital adrenal hyperplasia. *Psychoneuroendocrinology, 22,* 505–515.

Clark, M. M., & Galef, B. G. (1998). Effects of intrauterine position on the behavior and genital morphology of litter-bearing rodents. *Developmental Neuropsychology, 14,* 197–211.

Cohen-Bendahan, C. C. C., van de Beek, C., & Berenbaum, S. A. (2005). Prenatal sex hormone effects on child and adult sex-typed behavior: Methods and findings. *Neuroscience and Biobehavioral Reviews, 29,* 353–384.

Colcombe, S. J., Kramer, A. F., Erickson, K. I., Scalf, P., McAuley, E., Cohen, N. J., et al. (2004). Cardiovascular fitness, cortical plasticity, and aging. *Proceedings of the National Academy of Science, 101,* 3316–3321.

Goldapple, K., Segal, Z., Garson, C., Lau, M., Bieling, P., Kennedy, S., et al. (2004). Modulation of cortical-limbic pathways in major depression: Treatment-specific effects of cognitive behavior therapy. *Archives of General Psychiatry, 61,* 34–41.

Hampson, E., Rovet, J. F., & Altmann, D. (1998). Spatial reasoning in children with congenital adrenal hyperplasia due to 21-hydroxylase deficiency. *Developmental Neuropsychology, 14,* 299–320.

Hines, M., Brook, C., & Conway, G. S. (2004). Androgen and psychosexual development: Core gender identity, sexual orientation, and recalled gender role behavior in women and men with congenital adrenal hyperplasia (CAH). *Journal of Sex Research, 41,* 75–81.

Hines, M., Fane, B. A., Pasterski, V. L., Mathews, G. A., Conway, G. S., & Brook, C. (2003). Spatial abilities following prenatal androgen abnormality: Targeting and mental rotations performance in individuals with congenital adrenal hyperplasia. *Psychoneuroendocrinology, 28,* 1010–1026.

Hines, M., & Kaufman, F. (1994). Androgen and the development of human sex-typical behavior: Rough-and-tumble play and sex of preferred playmates in children with congenital adrenal hyperplasia (CAH). *Child Development, 65,* 1042–1053.

Leveroni, C. L., & Berenbaum, S. A. (1998). Early androgen effects on interest in infants: Evidence from children with congenital adrenal hyperplasia. *Developmental Neuropsychology, 14,* 321–340.

Low, K. S. D., Yoon, M., Roberts, B. W., & Rounds, J. (2005). The stability of vocational interests from early adolescence to middle adulthood: A quantitative review of longitudinal studies. *Psychological Bulletin, 131,* 713–737.

Maccoby, E. E. (1998). *The two sexes: Growing up apart, coming together.* Cambridge, MA: Harvard University Press.

Maguire, E. A., Spiers, H. J., Good, C. D., Hartley, T., Frackowiak, R. S., & Burgess, N. (2003). Navigation expertise and the human hippocampus: A structural brain imaging analysis. *Hippocampus, 13,* 250–259.

Meyer-Bahlburg, H. F. L. (1999). What causes low rates of child-bearing in congenital adrenal hyperplasia? *Journal of Clinical Endocrinology and Metabolism, 84,* 1844–1847.

Meyer-Bahlburg, H. F. L. (2001). Gender and sexuality in congenital adrenal hyperplasia. *Endocrinology and Metabolism Clinics of North America, 30,* 155–171.

Moore, C. L. (1992). The role of maternal stimulation in the development of sexual behavior and its neural basis. *Annals of the New York Academy of Sciences, 662,* 160–177.

Nordenström, A., Servin, A., Bohlin, G., Larsson, A., & Wedell, A. (2002). Sex-typed toy play behavior correlates with the degree of prenatal androgen exposure assessed by CYP21 genotype in girls with congenital adrenal hyperplasia. *Journal of Clinical Endocrinology and Metabolism, 87,* 5119–5124.

Paquette, V., Levesque, J., Mensour, B., Leroux, J. M., Beaudoin, G., Bourgouin, P., et al. (2003). "Change the mind and you change the brain": Effects of cognitive–behavioral therapy on the neural correlates of spider phobia. *NeuroImage*, *18*, 401–409.

Pasterski, V. L., Geffner, M. E., Brain, C., Hindmarsh, P., Brook, C., & Hines, M. (2005). Prenatal hormones and postnatal socialization by parents as determinants of male-typical toy play in girls with congenital adrenal hyperplasia. *Child Development, 76*, 264–278.

Pollak, S. D., Klorman, R., Thatcher, J. E., & Cicchetti, D. (2001). P3b reflects maltreated children's reactions to facial displays of emotion. *Psychophysiology, 38*, 267–274.

Resnick, S. M. (2006). Sex differences in regional brain structure and function. In P. W. Kaplan (Ed.), *Neurologic disease in women* (2nd ed., pp. 15–26). New York: Demos Medical.

Resnick, S. M., Berenbaum, S. A., Gottesman, I. I., & Bouchard, T. J. (1986). Early hormonal influences on cognitive functioning in congenital adrenal hyperplasia. *Developmental Psychology, 22*, 191–198.

Ruble, D. N., Martin, C. L., & Berenbaum, S. A. (2006). Gender development. In W. Damon & R. M. Lerner (Series Eds.) & N. Eisenberg (Vol. Ed.), *Handbook of child psychology: Vol. 3. Social, emotional, and personality development* (6th ed., pp. 858–932). New York: Wiley.

Servin, A., Nordenström, A., Larsson, A., & Bohlin, G. (2003). Prenatal androgens and gender-typed behavior: A study of girls with mild and severe forms of congenital adrenal hyperplasia. *Developmental Psychology, 39*, 440–450.

Wallen, K. (1996). Nature needs nurture: The interaction of hormonal and social influences on the development of behavioral sex differences in rhesus monkeys. *Hormones and Behavior, 30*, 364–378.

Wallen, K. (2005). Hormonal influences on sexually differentiated behavior in non-human primates. *Frontiers in Neuroendocrinology, 26*, 7–26.

Wood, W., & Eagly, A. H. (2002). A cross-cultural analysis of the behavior of women and men: Implications for the origins of sex differences. *Psychological Bulletin, 128*, 699–727.

Zucker, K. J., Bradley, S. J., Oliver, G., Blake, J., Fleming, S., & Hood, J. (1996). Psychosexual development of women with congenital adrenal hyperplasia. *Hormones and Behavior, 30*, 300–318.

12

SEX DIFFERENCES IN MIND: KEEPING SCIENCE DISTINCT FROM SOCIAL POLICY

SIMON BARON-COHEN

There are interesting differences between the average male and female mind. In using the word *average*, I am from the outset recognizing that such differences may have little to say about individuals. In addition, the differences are subtle and are to do with the relative proportions of different drives in the typical male and female mind. The field of sex differences in psychology in the 1960s and 1970s was so conflict-ridden as to make an open-minded debate about any potential role of biology contributing to psychological sex differences impossible. Those who explored the role of biology—even while acknowledging the importance of culture—found themselves accused of defending an essentialism that perpetuated inequalities between the sexes, and of oppression. Not a climate in which scientists can ask questions about mechanisms in nature. Today, the pendulum has settled sensibly in the middle of the nature–nurture debate, and scientists who care deeply about ending inequality and oppression can at the same time also talk freely about biological differences between the average male and female brain and mind.

Parts of this chapter were presented in 2005 at the Darwin Lecture Series, Cambridge University, Cambridge, England. The author was supported by the Medical Research Council and the Nancy Lurie Marks Family Foundation during the preparation of this work.

My own view is that the field of sex differences in mind needs to proceed in a fashion that is sensitive to this history of conflict by cautiously looking at the evidence and being careful not to overstate what can be concluded. Once again, the evidence says nothing about individuals. As we will see, the data actually require us to look at each individual on his or her own merits, as individuals may or may not be typical for their sex. In this chapter I first look at the evidence from scientific studies of sex differences in the mind. At the end of the chapter, in keeping with the theme of this edited collection, I consider the separate social policy issue of whether, as a society, we can achieve equal representation of women and men in science if we aim to do so.

SYSTEMIZING AND EMPATHIZING

Empathizing is the drive to identify another person's emotions and thoughts and to respond to these with an appropriate emotion. Empathizing allows you to *predict* a person's behavior and to care about how others feel. In this chapter, I review the evidence that, in general, females spontaneously empathize to a greater degree than do males. *Systemizing* is the drive to analyze the variables in a system to derive the underlying rules that govern its behavior. Systemizing also refers to the drive to construct systems. Systemizing allows one to predict the behavior of a system and to control it. I review the evidence that, on average, males spontaneously systemize to a greater degree than do females (Baron-Cohen, Wheelwright, Lawson, Griffin, & Hill, 2002).

Empathizing is close enough to the standard English definition to need little introduction, and I will return to it shortly. However, systemizing is a new concept and needs a little more definition. By a *system*, I mean something that takes inputs and delivers outputs. To systemize, one uses "if-then" (correlation) rules. The brain zooms in on a detail or parameter of the system and observes how this varies. That is, it treats a feature of a particular object or event as a variable. Alternatively, a person actively, or systematically, manipulates a given variable. One notes the effect(s) of operating on one single input in terms of its effects elsewhere in the system (the output). The key data structure used in systemizing is input–operation–output. If I do x, a changes to b. If z occurs, p changes to q. Systemizing therefore requires an exact eye for detail.

There are at least six main kinds of systems that the human brain can analyze or construct:

- technical systems (e.g., a computer, a musical instrument, a hammer);
- natural systems (e.g., a tide, a weather front, a plant);

- abstract systems (e.g., mathematics, a computer program, syntax);
- social systems (e.g., a political election, a legal system, a business);
- organizable systems (e.g., a taxonomy, a collection, a library); and
- motoric systems (e.g., a sports technique, a performance, a musical technique).

Systemizing is an inductive process. One watches what happens each time, gathering data about an event from repeated sampling, often quantifying differences in some variables within the event and observing their correlation with variation in outcome. After confirming a reliable pattern of association—that is, generating predictable results—one forms a rule about how a particular aspect of the system works. When an exception occurs, the rule is refined or revised. Otherwise, the rule is retained. Systemizing works for phenomena that are ultimately lawful, finite, and deterministic. The explanation is exact, and its truth value is testable (e.g., "The light went on because the switch was in the down position"). Systemizing is of almost no use for predicting moment-to-moment changes in a person's behavior. To predict human behavior, empathizing is required. Systemizing and empathizing are wholly different kinds of processes.

Empathizing involves the attribution of mental states to others and involves an appropriate affective response to the other's affective state. It not only includes what is sometimes called *theory of mind*, or mentalizing (Morton, Leslie, & Frith, 1995) but also encompasses the common English words *empathy* and *sympathy*. Although systemizing and empathizing are in one way similar because they are processes that allow us to make sense of events and make reliable predictions, they are in another way almost the opposite of each other. Empathizing involves an imaginative leap in the dark in the absence of complete data (e.g., "Maybe she didn't phone me because she was feeling hurt by my comment"). The causal explanation is at best a "maybe," and its truth may never be provable. Systemizing is our most powerful way of understanding and predicting the law-governed inanimate universe. Empathizing is our most powerful way of understanding and predicting the social world. Ultimately, empathizing and systemizing depend on separate, independent regions in the human brain.

THE MAIN BRAIN TYPES

In this chapter, I argue that systemizing and empathizing are two key dimensions that define the male and female brain. We all have both systemizing and empathizing skills. One can envisage five broad types of brain, as Table 12.1 shows. This chapter concerns primarily those on the extreme

TABLE 12.1
The Main Brain Types

Profile	Shorthand equation	Type of brain
Individuals in whom empathizing is more developed than systemizing.	E > S	"Female" (or Type E)
Individuals in whom systemizing is more developed than empathizing.	S > E	"Male" (or Type S)
Individuals in whom systemizing and empathizing are both equally developed.	S = E	"Balanced" (or Type B)
Individuals in whom systemizing is hyperdeveloped while empathizing is hypodeveloped. (Individuals on the autistic spectrum have been found to fit this profile.)	S >> E	Extreme male brain
Individuals who have hyperdeveloped empathizing skills, while their systemizing is hypodeveloped.	E >> S	Extreme female brain (postulated)

male brain end of the spectrum. Individuals who have this psychological profile may be talented systemizers, but they are often, at the same time, "mind-blind" (Baron-Cohen, 1995). The evidence reviewed here suggests that not all men have the male brain and not all women have the female brain. Expressed differently, some women have the male brain, and some men have the female brain. My central claim here is that more males than females have a brain of Type S, and more females than males have a brain of Type E. Type S means the person has the profile of S > E (their systemizing is stronger than their empathy), and Type E means the person has the opposite profile (E > S; see Figure 12.1). Next, I review the evidence supporting these profiles. In the final section of this chapter, I highlight the role of culture and biology in these sex differences.

The Female Brain: Empathizing

What is the evidence for female superiority in empathizing? In the studies summarized here, sex differences of a small but statistically significant magnitude have been found.

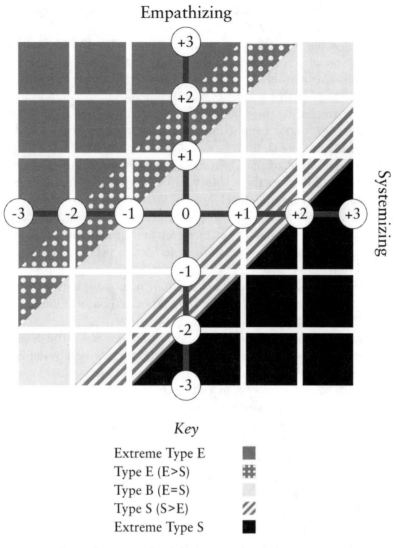

Key

Extreme Type E ▪

Type E (E>S) ✻

Type B (E=S) ▪

Type S (S>E) ▨

Extreme Type S ▪

Axes show standard deviations from the mean

Figure 12.1. A model of the different brain types.

- *Sharing and turn taking.* On average, girls show more concern for fairness, whereas boys share less. In one study, boys showed 50 times greater competition compared with girls, whereas girls showed 20 times greater turn taking compared with boys (Charlesworth & Dzur, 1987).
- *Rough-and-tumble play or "roughhousing"* (e.g., wrestling, mock fighting, etc.). Boys show more of this than do girls. Although such activity is often playful, it can hurt or be intrusive. Lower

empathizing levels are necessary to engage in rough-and-tumble play (Maccoby, 1998).

- *Responding empathically to the distress of other people.* Girls from the age of 1 year show greater concern for others through sad looks, sympathetic vocalizations, and comforting actions as compared with boys. Also, more women than men report frequently sharing the emotional distress of their friends. Women also show more comforting, even to strangers, than men do (Hoffman, 1977).

- *Using a "theory of mind."* As early as 3 years of age, little girls are ahead of boys in their ability to infer what people might be thinking or intending (Happe, 1995).

- *Sensitivity to facial expressions.* Women are better at decoding nonverbal communication, picking up subtle nuances from tone of voice or facial expression, or judging a person's character (Davis, 1994).

- *Empathy.* Women score higher than men on questionnaires designed to measure empathic response (Baron-Cohen & Wheelwright, 2004).

- *Values in relationships.* More women than men value the development of altruistic, reciprocal relationships, which by definition require empathizing. In contrast, more men value power, politics, and competition (Ahlgren & Johnson, 1979). Girls are more likely to endorse cooperative items on a questionnaire and to rate the establishment of intimacy as more important than the establishment of dominance. In contrast, boys are more likely than girls to endorse competitive items and to rate social status as more important than intimacy (Knight & Chao, 1989).

- *Disorders of empathy.* Disorders such as psychopathic personality disorder or conduct disorder are far more common among males (Blair, 1995; Dodge, 1980).

- *Aggression.* Even in normal quantities, this can only occur with reduced empathizing. Here again, there is a clear sex difference. Males tend to show far more direct aggression (e.g., pushing, hitting, punching, etc.), whereas females tend to show more indirect (i.e., relational, covert) aggression (e.g., gossip, exclusion, cutting remarks, etc.). Direct aggression may require an even lower level of empathy than indirect aggression. Indirect aggression needs better mind-reading skills than does direct aggression because its impact is strategic (Crick & Grotpeter, 1995).

- *Murder.* This is the ultimate example of a lack of empathy. Daly and Wilson (1988) analyzed homicide records dating back over

700 years, from a range of different societies, and found that male-on-male homicide was 30 to 40 times more frequent than female-on-female homicide.

- *Establishing a "dominance hierarchy."* Males are quicker to establish such hierarchies. This in part reflects their lower empathizing skills because often a hierarchy is established by one person pushing others around to become the leader (Strayer, 1980).

- *Language style.* Girls' speech is more cooperative, reciprocal, and collaborative. In concrete terms, this is also reflected in girls being able to continue a conversational exchange with a partner for a longer period. When girls disagree, they are more likely to express their different opinion sensitively, in the form of a question rather than an assertion. Boys' talk is more "single-voiced discourse"; that is, the speaker presents only his own perspective. The female speech style is more "double-voiced discourse"; girls spend more time negotiating with their partner, trying to take the other person's wishes into account (Smith, 1985).

- *Talk about emotions.* Women's conversations involve much more talk about feelings, whereas men's conversations tend to be more object or activity focused (Tannen, 1990).

- *Parenting style.* Fathers are less likely than mothers to hold their infants in a face-to-face position. Mothers are more likely to follow through the child's choice of topic in play, whereas fathers are more likely to impose their own topic. Also, mothers fine-tune their speech more often to match their children's understanding (Power, 1985).

- *Face preference and eye contact.* From birth, females look longer at faces, particularly at people's eyes, whereas males are more likely to look at inanimate objects (Connellan, Baron-Cohen, Wheelwright, Batki, & Ahluwalia, 2000).

- *Language ability.* Females have been shown to have better language ability than males. It seems likely that good empathizing would promote language development (Baron-Cohen, Baldwin, & Crowson, 1997), and vice versa, so these factors may not be independent.

The Male Brain: Systemizing

The relevant domains to explore for evidence of systemizing include any fields that are in principle rule-governed. Thus, chess and football are good examples of systems, but faces and conversations are not. As noted previously, systemizing involves monitoring three elements: input, operation, and output. The operation is what was done or what happened to the input to produce the output. What is the evidence for a stronger drive to systemize in males?

- *Toy preferences*. Boys are more interested than girls in toy ve-hicles, weapons, building blocks, and mechanical toys, all of which are open to being "systemized" (Jennings, 1977).
- *Adult occupational choices*. Some occupations are almost entirely male. These include metalworking, weapon making, manufac-ture of musical instruments, and the construction industries, such as boat building. The focus of these occupations is on cre-ating systems (Geary, 1998).
- *Math, physics, and engineering*. These disciplines all require high systemizing and are largely male dominated. The Scholastic Aptitude—Mathematics Test (SAT–M) is the mathematics part of the test administered nationally to college applicants in the United States. Males on average score 50 points higher than females on this test (Benbow, 1988). Considering only indi-viduals who score above 700, the sex ratio is 13:1 (men to women; Geary, 1996).
- *Constructional abilities*. On average, men score higher than women in an assembly task in which people are asked to put together a three-dimensional (3-D) mechanical apparatus. Boys are also better at constructing block buildings from two-dimensional blue-prints. Lego bricks can be combined and recombined into an infinite number of systems. Boys show more interest than girls in playing with Lego. Boys as young as 3 years of age are also faster at copying 3-D models of outsized Lego pieces. Older boys, from the age of 9 years, are better than girls at imagining what a 3-D object will look like if it is laid out flat. Boys are also better at constructing a 3-D structure from just an aerial and frontal view in a picture (Kimura, 1999).
- *The water level task*. Originally devised by the Swiss child psy-chologist Jean Piaget, the water level task involves a bottle that is tipped at an angle. Individuals are asked to predict the water level. Women more often draw the water level aligned with the tilt of the bottle and not horizontal, as is correct (Wittig & Allen, 1984).
- *The rod and frame test*. If a person's judgment of vertical is influ-enced by the tilt of the frame, he or she is said to be *field depen-dent*; that is, the person's judgment is easily swayed by extrane-ous input in the surrounding context. If a person is not influenced by the tilt of the frame, he or she is said to be *field independent*. Most studies indicate that females are more field dependent; that is, women are relatively more distracted by

contextual cues, and they tend not to consider each variable within a system separately. They are more likely than men to state erroneously that a rod is upright if it is aligned with its frame (Witkin et al., 1954).

- *Good attention to relevant detail.* This is a general feature of systemizing and is clearly a necessary part of it. Attention to relevant detail is superior in males. One measure of this is the embedded figures test. On average, males are quicker and more accurate in locating a target object from a larger, complex pattern (Elliot, 1961). Males, on average, are also better at detecting a particular feature (static or moving) than are women (Voyer, Voyer, & Bryden, 1995).

- *The mental rotation test.* This test provides another example in which males are quicker and more accurate than females. This test involves systemizing because it is necessary to treat each feature in a display as a variable that can be transformed (e.g., rotated) and then predict the output, or how it will appear after transformation (Collins & Kimura, 1997).

- *Reading maps.* This is another everyday test of systemizing, because features from 3-D input must be transformed to a two-dimensional representation. In general, boys perform at a higher level than girls in map reading. Men can also learn a route by looking at a map in fewer trials than women, and they are more successful at correctly recalling greater detail about direction and distance. This observation suggests that men treat features in the map as variables that can be transformed into three dimensions. When children are asked to make a map of an area that they have only visited once, boys' maps have a more accurate layout of the features in the environment. Girls make more serious errors in the map location of important landmarks. Boys tend to emphasize routes or roads, whereas girls tend to emphasize specific landmarks (the corner shop, the park, etc.). These strategies of using directional cues versus using landmark cues have been widely studied. The directional strategy represents an approach to understanding space as a geometric system. Similarly, the focus on roads or routes is an example of considering space in terms of another system, in this case a transportation system (Galea & Kimura, 1993).

- *Motoric systems.* When people are asked to throw or catch moving objects (target-directed tasks), such as playing darts or intercepting balls flung from a launcher, males tend to perform better than females. In addition, on average men are more accurate than women in their ability to judge which of two moving objects is traveling faster (Schiff & Oldak, 1990).

- *Organizable systems.* People in the Aguaruna tribe of northern Peru were asked to classify a hundred or more examples of local specimens into related species. Men's classification systems included more subcategories (i.e., they introduced greater differentiation) and were more consistent among individuals. It is interesting that the criteria the Aguaruna men used to decide which animals belonged together more closely resembled the taxonomic criteria used by Western (mostly male) biologists (Atran, 1994). Classification and organization involve systemizing because categories are predictive. With more fine-grained categories, a system will provide more accurate predictions.
- *The Systemizing Quotient.* This is a questionnaire that has been tested among adults in the general population. It includes 40 items that ask about a subject's level of interest in a range of different systems that exist in the environment, including technical, abstract, and natural systems. Males score higher than females on this measure (Baron-Cohen, Richler, Bisarya, Gurunathan, & Wheelwright, 2003).
- *Mechanics.* The Physical Prediction Questionnaire is based on an established method for selecting applicants to study engineering. The task involves predicting which direction levers will move when an internal mechanism of cog wheels and pulleys is engaged. Men score significantly higher on this test compared with women (Lawson, Baron-Cohen, & Wheelwright, 2004).

CULTURE AND BIOLOGY

At age 1 year, boys strongly prefer to watch a video of cars going past, an example of predictable mechanical systems, than to watch a film showing a human face. Little girls show the opposite preference. Young girls also demonstrate more eye contact than do boys at age 1 year (Lutchmaya & Baron-Cohen, 2002). Some investigators argue that, even by this age, socialization may have caused these sex differences. Although evidence exists for differential socialization contributing to sex differences, this is unlikely to be a sufficient explanation. Connellan and colleagues showed that among 1-day-old babies, boys look longer at a mechanical mobile, which is a system with predictable laws of motion, than at a person's face, an object that is next to impossible to systemize; 1-day-old girls show the opposite profile (Connellan et al., 2000). These sex differences are therefore present very early in life. This raises the possibility that, although culture and socialization may partly determine the development of a male brain with a stronger interest in systems or a female brain with a stronger interest in empathy, biology may also partly determine this. There is ample evidence to support both cultural determinism and biological determinism (Eagly, 1987; Gouchie & Kimura, 1991). For example, the amount of time a 1-year-old child maintains eye contact is inversely related to the prenatal level of testosterone (Lutchmaya, Baron-Cohen, & Raggatt, 2002). The evidence for the biological basis of sex differences in the mind is reviewed elsewhere (Baron-Cohen, 2003).

CONCLUSIONS AND IMPLICATIONS FOR WOMEN IN SCIENCE

The previously discussed evidence suggests that the male brain is characterized by Type S (where S > E), the female brain by Type E (where E > S). What are the implications of such research for our view of women in science? This research suggests that we should not expect the sex ratio in occupations such as math or physics to ever be 50–50 if we leave the workplace to simply reflect the numbers of applicants of each sex who are drawn to such fields. The assumption here is that just as if you leave toys out on the carpet and film if boys and girls spontaneously choose to play with the same or different toys, you find that more boys play with the toys that involve systemizing (e.g., constructional or mechanical toys) and more girls play with the toys that involve empathizing (e.g., caring for dolls), so it might be that we will always see more males spontaneously choosing to apply to work in fields that involve systemizing (science, engineering, auto mechanics, etc.) and more females spontaneously choosing to work in fields that involve empathy (telephone help lines for those with mental health crises, e.g., the "Samaritans"). Of course, the question of how one determines if a person's choice is spontaneous or determined by cultural or biological factors is extremely hard to pin down. The study of newborn babies which found that more newborn boys spontaneously look longer at a mechanical mobile, and more newborn girls spontaneously look longer at a human face, suggests biology plays one part in leading to this "bias" in attention to things rather than emotions (in boys) and vice versa (in girls). Yet this is not to minimize the major role that culture also plays in amplifying such partly innate differences as the child grows up.

A key argument, reflected in the title of this chapter, is that we should separate the scientific question ("Are there sex differences in mind?") from the social policy agenda ("How can we achieve equal representation of women in science, or in any field?"). This is because they can be considered separately. If we want a particular field to have an equal representation of men and women, which may be ethically desirable in terms of equality of opportunity, equality of status, equality of income, or ensuring balance in the workplace, then we need to put in place social policies that will bring about that outcome. In other fields, it will not be necessary to intervene with policy. Medicine is a good example of a science in which female applicants now outnumber male applicants, probably because it is a science that favors the Type B brain (S = E, or good systemizing together with good empathy), and Type B is actually more common among females. Yet math and physics may have little or no role for empathy, and so favor the Type S brain that is more common in males.

Finally, the research teaches us that there is no scientific justification for stereotyping, because none of the studies allow one to predict an individual's aptitudes or interests on the basis of their sex. This is because—at risk of repetition—they only capture differences between groups on average. Individuals are just that: They may be typical or atypical for their group (their sex). The applicant for the job in your science department may be a woman with a more typically "male" brain or may be a man with a more typically "female" brain. Which means that to prejudge an individual on the basis of sex is, as the word *prejudge* suggests, mere prejudice. We need to look at applicants on the basis of who they are as individuals, not on the basis of their sex, when judging their aptitude.

REFERENCES

Ahlgren, A., & Johnson, D. W. (1979). Sex differences in cooperative and competitive attitudes from the 2nd through the 12th grades. *Developmental Psychology, 15,* 45–49.

Atran, S. (1994). Core domains versus scientific theories: Evidence from systematics and Itzaj-Maya folkbiology. In L. A. Hirschfeld & S. A. Gelman (Eds.), *Mapping the mind: Domain specificity in cognition and culture* (pp. 316–340). Cambridge, England: Cambridge University Press.

Baron-Cohen, S. (1995). *Mindblindness: An essay on autism and theory of mind.* Boston: MIT Press.

Baron-Cohen, S. (2003). *The essential difference: Men, women, and the extreme male brain.* New York: Basic Books.

Baron-Cohen, S., Baldwin, D. A., & Crowson, M. (1997). Do children with autism use the speaker's direction of gaze strategy to crack the code of language? *Child Development, 68,* 48–57.

Baron-Cohen, S., Richler, J., Bisarya, D., Gurunathan, N., & Wheelwright, S. (2003). The Systemising Quotient (SQ): An investigation of adults with Asperger syndrome or high functioning autism and normal sex differences. *Philosophical Transactions of the Royal Society, Series B, 358,* 361–374.

Baron-Cohen, S., & Wheelwright, S. (2004). The Empathy Quotient (EQ): An investigation of adults with Asperger syndrome or high functioning autism, and normal sex differences. *Journal of Autism and Developmental Disorders, 34,* 163–175.

Baron-Cohen, S., Wheelwright, S., Lawson, J., Griffin, R., & Hill, J. (2002). The exact mind: Empathising and systemising in autism spectrum conditions. In

U. Goswami (Ed.), *Handbook of cognitive development* (pp. 491–508). Oxford, England: Blackwell.

Benbow, C. P. (1988). Sex differences in mathematical reasoning ability in intellectually talented preadolescents: Their nature, effects, and possible causes. *Behavioral and Brain Sciences, 11,* 169–232.

Blair, R. J. (1995). A cognitive developmental approach to morality: Investigating the psychopath. *Cognition, 57,* 1–29.

Charlesworth, W. R., & Dzur, C. (1987). Gender comparisons of preschoolers' behavior and resource utilization in group problem solving. *Child Development, 58,* 191–200.

Collins, D. W., & Kimura, D. (1997). A large sex difference on a two-dimensional mental rotation task. *Behavioral Neuroscience, 111,* 845–849.

Connellan, J., Baron-Cohen, S., Wheelwright, S., Batki, A., & Ahluwalia, J. (2000). Sex differences in human neonatal social perception. *Infant Behavior and Development, 23,* 113–118.

Crick, N. R., & Grotpeter, J. K. (1995). Relational aggression, gender, and social–psychological adjustment. *Child Development, 66,* 710–722.

Daly, M., & Wilson, M. (1988). *Homicide.* New York: Aldine de Gruyter.

Davis, M. H. (1994). *Empathy: A social psychological approach.* In J. Harvey (Ed.), *Brown & Benchmark social psychology series.* Boulder, CO: Westview Press.

Dodge, K. A. (1980). Social cognition and children's aggressive behavior. *Child Development, 51,* 162–170.

Eagly, A. H. (1987). *Sex differences in social behavior: A social-role interpretation.* Hillsdale, NJ: Erlbaum.

Elliot, R. (1961). Interrelationship among measures of field dependence, ability, and personality traits. *Journal of Abnormal and Social Psychology, 63,* 27–36.

Galea, L. A. M., & Kimura, D. (1993). Sex differences in route-learning. *Personality and Individual Differences, 14,* 53–65.

Geary, D. C. (1996). Sexual selection and sex differences in mathematical abilities. *Behavioral and Brain Sciences, 19,* 229–284.

Geary, D. C. (1998). *Male, female: The evolution of human sex differences.* Washington, DC: American Psychological Association.

Gouchie, C., & Kimura, D. (1991). The relationship between testosterone levels and cognitive ability patterns. *Psychoneuroendocrinology, 16,* 323–334.

Happe, F. G. (1995). The role of age and verbal ability in the theory of mind task performance of subjects with autism. *Child Development, 66,* 843–855.

Hoffman, M. L. (1977). Sex differences in empathy and related behaviors. *Psychological Bulletin, 84,* 712–722.

Jennings, K. D. (1977). People versus object orientation in preschool children: Do sex differences really occur? *Journal of Genetic Psychology, 131,* 65–73.

Kimura, D. (1999). *Sex and cognition.* Cambridge, MA: MIT Press.

Knight, G. P., & Chao, C.-C. (1989). Gender differences in the cooperative, competitive, and individualistic social values of children. *Motivation and Emotion, 13*, 125–141.

Lawson, J., Baron-Cohen, S., & Wheelwright, S. (2004). Empathizing and systemizing in adults with and without Asperger syndrome. *Journal of Autism and Development Disorders, 34*, 301–310.

Lutchmaya, S., & Baron-Cohen, S. (2002). Human sex differences in social and nonsocial looking preferences at 12 months of age. *Infant Behavior and Development, 25*, 319–325.

Lutchmaya, S., Baron-Cohen, S., & Raggatt, P. (2002). Foetal testosterone and eye contact in 12-month-old infants. *Infant Behaviour and Development, 25*, 327–335.

Maccoby, E. E. (1998). *The two sexes: Growing up apart, coming together*. Cambridge, MA: Belknap Press/Harvard University Press.

Morton, J., Leslie, A., & Frith, U. (1995). The cognitive basis of a biological disorder: Autism. *New Scientist, 14*, 434–438.

Power, T. G. (1985). Mother– and father–infant play: A developmental analysis. *Child Development, 56*, 1514–1524.

Schiff, W., & Oldak, R. (1990). Accuracy of judging time to arrival: Effects of modality, trajectory, and gender. *Journal of Experimental Psychology: Human Perception and Performance, 16*, 303–316.

Smith, P. M. (1985). *Language, the sexes, and society*. Oxford, England: Blackwell.

Strayer, F. F. (1980). Child ethology and the study of preschool social relations. In H. C. Foot, A. J. Chapman, & J. R. Smith (Eds.), *Friendship and social relations in children* (pp. 235–265). Chichester, England: Wiley.

Tannen, D. (1990). *You just don't understand: Women and men in conversation*. New York: William Morrow.

Voyer, D., Voyer, S., & Bryden, M. (1995). Magnitude of sex differences in spatial abilities: A meta-analysis and consideration of critical variables. *Psychological Bulletin, 117*, 250–270.

Witkin, H. A., Lewis, H. B., Hertzman, M., Machover, K., Bretnall, K., Meissner, P., & Wapner, S. (1954). *Personality through perception*. New York: Harper.

Wittig, M. A., & Allen, M. J. (1984). Measurement of adult performance on Piaget's water horizontality task. *Intelligence, 8*, 305–313.

13

AN EVOLUTIONARY PERSPECTIVE ON SEX DIFFERENCES IN MATHEMATICS AND THE SCIENCES

DAVID C. GEARY

Sex differences in mathematics and the sciences can be expressed in terms of grades, scores on standardized tests, and creative contributions. When sex differences are found, they are related to factors that influence engagement in mathematical or scientific endeavors and to cognitive (e.g., spatial ability) and intellectual (e.g., working memory) competencies that support this engagement. Engagement in these endeavors can be related to social factors and personal interest, and competencies that facilitate learning can be influenced by biology, experience, or some combination. Obviously, the issues are multifaceted, and any single chapter is necessarily limited in scope. I ask you to consider the possibility that sex differences in certain cognitive abilities and personal interests contribute to long-term engagement in mathematical and scientific endeavors and to ease of learning some aspects of mathematics and the sciences. I approach these issues from an evolutionary perspective. This is an especially complicated approach because the knowledge bases, technical skills, and conceptual insights in mathematics and the sciences arise from a poorly understood interaction between inherent cognitive and motivational biases and culture-specific educational goals and opportunities. Nonetheless, a comprehensive analysis of sex differences in mathematics and the sciences requires consideration of potentially evolved but indirect influences. I first review sex differences in school and in terms of exceptional accomplishment, and then provide a primer on the mechanisms that result in the evolution of sex differences. In the third section, I provide an overview of the evidence for these mechanisms in humans, and in the final section I outline potential links between evolved sex differences and sex differences in mathematics and the sciences.

OUTCOMES IN MATHEMATICS AND THE SCIENCES

School

Girls and women typically earn higher grades in mathematics and science classes (Willingham & Cole, 1997) and have made substantial gains in baccalaureate and doctoral degrees during the past 4 decades (Hill & Johnson, 2004). During this same epoch, boys and men posted higher average scores and outnumbered girls and women at the high end on standardized mathematics and science tests that include novel (i.e., not directly taught in school) problems, such as the Scholastic Achievement Test—Mathematics (SAT–M; Halpern, chap. 9, this volume; Hedges & Nowell, 1995). With the SAT–M and related tests, the difference in average scores and the number of boys and men at the high end have not changed substantially since the 1960s (Stumpf & Stanley, 1998). Long-term studies have shown that performance on the SAT–M predicts later success in mathematics, the sciences, and engineering, as does performance on standardized measures of spatial ability (Lubinski & Benbow, chap. 6, this volume). These standardized measures capture important dimensions of what it takes to excel in these fields (Nuttall, Casey, & Pezaris, 2005).

Exceptional Accomplishment

Historical bursts of creative activity (e.g., Renaissance) tended to emerge in wealthier cultures with mores that did not restrict individual freedom and rewarded creative expression, and mathematics and the sciences were no exception. Studies of extraordinary accomplishments suggest they tend to be generated by individuals who exhibit a combination of traits, including high general intelligence, creativity, an extended period of preparation (~10 years), potentially domain-specific abilities (e.g., spatial visualization or language), ambition, and sustained productivity (Simonton, 1999). Individuals at the extreme of all of these dimensions are the most likely to make eminent contributions. Historically, the ratio of men to women who made influential and long-lasting contributions in mathematics and the sciences was as high as 50 to 1 (Murray, 2003); the ratio was 10 to 1 in literature. Because accomplishment at this level requires the just listed mix of cultural opportunity and intraindividual traits, a sex difference on only a single dimension could result in a substantial male advantage, even with no differences for other relevant traits. For instance, the restricted opportunities of women during the period when the foundations for modern mathematics and the sciences were built would have resulted in many more men than women building this foundation. Yet are restricted opportunities a sufficient explanation? The other potential traits contributing to exceptional accomplishment have not been fully considered and thus the issue is unresolved.

SEXUAL SELECTION

Evolution

Darwin's (1871) sexual selection describes the most common dynamics associated with the evolution of sex differences, that is, male–male (*intrasexual*) competition for mates and female choice (*intersexual choice*) of mating partners. The most common result is the elaboration of traits that facilitate competition and choice. These dynamics turn on the degree to which each sex invests in parenting and on differences in the potential rate of reproduction (Clutton-Brock & Vincent, 1991; Trivers, 1972). The basic cross-species pattern is that the sex with the slower potential rate of reproduction (typically females, because of gestation time) invests more in parenting, is selective in mate choices, and exhibits less competition over mates. The sex with the faster potential rate of reproduction (typically males) invests less in parenting, is less selective in mate choices, and exhibits more intense competition. The accompanying prediction of more female investment in parenting and more intense intrasexual competition in males is found in more than 95% of mammalian species. Critically, the predicted reversal of sex differences is found for species in which males invest heavily in parenting (e.g., males incubate eggs). Females of these species are predictably larger, more colorful, and more pugnacious than males (Eens & Pinxten, 2000).

Sexual selection can also influence the evolution of brain and cognitive traits, as illustrated by studies of voles (*Microtus*; Gaulin, 1992). Male meadow voles (*M. pennsylvanicus*) are polygynous and compete by searching for and attempting to mate with females dispersed throughout the habitat. Males of monogamous prairie (*M. ochrogaster*) and pine voles (*M. pinetorum*) do not search for additional mates, once paired. For meadow voles but not prairie and pine voles, intrasexual competition favors males that court many females, which is possible only through expansion of the home range. This form of male–male competition should result in larger home ranges and better spatial abilities for male than female meadow voles and no sex difference for prairie or pine voles, which is exactly what is observed.

Development

Hormones

The expression of evolved sex differences will be influenced by exposure to sex hormones, especially androgens (e.g., testosterone; Morris, Jordan, & Breedlove, 2004). Androgens influence sex differences in cognition and behavioral biases through early prenatal organization of associated brain areas and through activation of these areas with postnatal exposure. For male meadow voles, testosterone increases significantly during the breeding season and spurs the increased activity levels needed to expand home ranges. The influences of sex hormones are, however, complex and often interact with genetic sex, physical health, and social and ecological context (Arnold et al., 2004). For evolved behaviors that require an extended period of learning, testosterone may act to increase these behaviors during development but does not in and of itself result in adult-level competencies; these require practice during the developmental period.

Within-Sex Variation

Consideration of differences within each sex is important, because it is individuals at the extreme of many traits who are most likely to make eminent contributions to academic fields. For nonhuman species, sexual selection can exaggerate within-sex variation in the traits that influence competition and choice (Pomiankowski & Møller, 1995). Because of a combination of genetic and hormonal influences, these traits are often condition-dependent; that is, their expression is heavily influenced by individual health and by social and ecological conditions during development and in adulthood. The resulting bias for extreme expression of these traits is maintained because their expression in fit individuals eliminates potential competitors that cannot express the traits to the same extreme. The basic point is that the sex that experiences more intense intrasexual competition or more intense vetting by the other sex will tend to show greater within-sex variation on many traits.

HUMAN SEX DIFFERENCES

Sexual Selection

Across species, phsycial male–male competition is associated with larger males than females, so sex differences in physical size are one indicator of sex differences in intrasexual competition. Thus, the human sex difference in physical size, pattern of physical development, and other traits strongly suggests sexual selection has contributed to some currently observed sex differences. Men's investment in parenting provides an important twist,

DAVID C. GEARY

however, and results in female–female competition and male choice, in addition to male–male competition and female choice. The details are beyond the scope of this chapter (Geary, 1998), but aspects of male–male competition are relevant.

In traditional societies this competition includes coordinated group-level conflict for control of ecologically rich territories and for social and political influence, which is often manifested as warfare and political manipulation (Keeley, 1996). Maintaining the groups' territorial borders, tribal warfare, and large game hunting within these borders—almost exclusively male activities (Murdock, 1981)—involve fluid movement in large ranges and should thus result in an evolved sex difference in the spatial abilities that support navigation. Also associated with hunting and warfare is the construction of weapons and other tools (also primarily a male activity), among other "folk" physical competencies, as described below.

Folk Domains

The evolved function of behavior is to allow individuals to gain control of the types of resource or avoid the types of threats that tended to covary with survival or reproductive prospects during the species' evolution (Geary, 2005). Most evolutionarily relevant resources or threats fall into three categories—social, biological, and physical—and the associated traits coalesce around the respective domains of folk psychology, folk biology, and folk physics (Hirschfeld & Gelman, 1994). Folk psychology is composed of the systems that enable people to negotiate social demands and includes knowledge related to the self, dyadic relationships, and group-level interactions. Human folk biology supports the categorizing of plants and animals in the local ecology, and knowledge of growth patterns and behavior that facilitate hunting and other activities involved in using these species as food and medicine (Atran, 1998). Folk physics includes the systems that enable navigating in three-dimensional space, mentally representing this space (e.g., group's territory) and for using physical materials for tool making (Pinker, 1997; Shepard, 1994). There may also be evolved systems for representing small quantities and for manipulating these representations by means of counting and simple arithmetic (Geary, 1995). In any case, the knowledge and behavioral skills that compose folk systems will emerge from an interaction of inherent constraints and experiences during development.

DEVELOPMENT AND INTERESTS

The developmental period is twice as long in humans as in related species. This reproductive delay may reflect the evolution of increased plasticity in brain, cognitive, and behavioral traits that compose folk domains, and through this an enhanced ability to adapt to novel conditions. Children's social play and exploration of the environment allow them to practice and refine the competencies that compose folk domains and that result in competitive advantage in adulthood (Bjorklund & Pellegrini, 2002). Sex differences in developmental activities are predicted to mirror sex differences in patterns of intrasexual competition, intersexual choice, and parental investment. These sex differences are predicted to be influenced by pre- and postnatal exposure to sex hormones, but the developing competencies will emerge from interactions between hormone-influenced biases in child-initiated activities and the specifics of the niches in which the children grow up. Indeed, many of the predicted sex differences in social biases (e.g., desire for dominance) and play activities (e.g., rough-and-tumble play) are found in many cultures (Whiting & Edwards, 1988) and are influenced by exposure to sex hormones (Cohen-Bendahan, van de Beek, & Berenbaum, 2005).

Child-initiated activities related to certain spatial and mechanical abilities may provide a link to later emerging sex differences in some competencies related to mathematics and the sciences. For instance, in industrial and traditional societies, boys' play ranges are $1^1/_2$ to 3 times the size of girls' play ranges, and boys manipulate the ecology within this range more frequently and in more complex ways (e.g., build forts; Matthews, 1992). These activity differences create a gap between boys and girls in the ability to mentally visualize and remember the geometric features of large-scale space. The sex difference in the size of the play range appears to result from a combination of parental restrictions on girls' exploration and child-initiated preferences that are related to prenatal exposure to androgens (Resnick, Berenbaum, Gottesman, & Bouchard, 1986; Whiting & Edwards, 1988), although the latter relation is complex and not fully understood (Hines et al., 2003).

The majority of studies of infants and toddlers have not found sex differences in some components of folk physics, such as discriminating mechanical from human motion (Spelke, 2005). Nonetheless, there are consistent sex differences in preschoolers' and older children's play preferences, whereby boys engage in more play with objects and girls engage in more family and parenting play (Cohen-Bendahan et al., 2005). The former differences are specific to certain types of objects and activities. Girls engage in more play with puzzles, markers, and so on, and boys engage in a more restricted category of play with inanimate mechanical objects (e.g., toy cars) and construction play that involves building. Golombok and Rust (1993) obtained from mothers information about the play activities of 2,330 preschoolers from three nations, and then cross-validated these with separate reports from some of the children's preschool teachers. The overall sex difference in early play activities was very large, and for individual activities was largest for play with toy vehicles and toy weapons, which favored boys, and risk avoidance and pretending to be a female character, which favored girls. These findings were confirmed in a study of 3,990 preschool twins and their siblings (Iervolino, Hines, Golombok, Rust, & Plomin, 2005) and revealed that about one third and one half of the respective individual differences in play preferences among boys and girls were related to genetic influences. Family influences were much stronger for boys than for girls, and influences common to twins, possibly prenatal exposure to sex hormones, were found for both sexes.

Play preferences common in boys are also found for girls with congenital adrenal hyperplasia (CAH), which results in excess prenatal exposure to androgens. For 3- to 8-year olds, Berenbaum and Hines (1992) reported that four out of five girls with CAH played with blocks, toy cars, and other "boys' toys" more often than their unaffected sisters or cousins. Three to 4 years later, the difference in preference for "boy play" had increased. Within the normal range of early hormone exposure, Hines et al. (2002) found that higher maternal testosterone levels during pregnancy were associated with more masculine play for their preschool daughters but not sons. For 3-year-olds, Gredlein and Bjorklund (2005) found engagement in a boy-typical form of object-oriented play was associated with skilled tool use during problem solving for boys but not girls. For 18-month-olds, Chen and Siegler (2000) found small to moderate advantages for boys for transfer of tool use from one setting to an analogous setting, in the consistency of tool use across settings, and in successful use of tools.

Regarding folk biology, in traditional societies women have more knowledge about plants and men about animals (Atran, 1998). However, these differences could simply reflect time spent in foraging and hunting, and not an inherent sex difference in attentional biases and preferences. The question remains to be answered, although there is some evidence that boys attend to wild and potentially dangerous animals more than girls do and engage in more play hunting in these societies (Blurton Jones, Hawkes, & O'Connell, 1997). Either way, any sex differences in cognitive competencies and interest in folk biological domains are predicted to be more nuanced than those found in folk physical domains, because both sexes obtain resources from the ecology in traditional societies and almost certainly did so throughout human evolution.

EVOLUTION, MATHEMATICS, AND THE SCIENCES

We are only beginning to link the learning of academic knowledge to the brain and cognitive systems that compose evolved folk domains, and the following is thus preliminary (see Geary, 2005). I first illustrate several potential links between folk knowledge and academic learning in the sciences and mathematics, and then relate these to sex differences.

Folk Knowledge, Science, and Mathematics

The Sciences

The evolved but naïve folk understanding of certain physical phenomena may have influenced the initial emergence of physics as a domain of conscious intellectual activity. For instance, when asked about the forces acting on a thrown baseball, most people believe there are forces propelling it forward, something like an invisible engine, and propelling it downward. The downward force is gravity, but there is no force propelling it forward, once the ball leaves the player's hand (Clement, 1982). The concept of a forward force, called *impetus*, is similar to pre-Newtonian beliefs about motion; starting an object in motion imparts to the object an internal force that keeps it in motion until this impetus dissipates. Even though adults often describe the correct trajectory of a thrown object, reflecting their implicit folk knowledge, their explicit explanations are often scientifically naive.

To correct such attributional errors, careful observation, use of the scientific method, and inductive and deductive reasoning are needed. Newton (1687/1995, p. 13) said as much in the *Principia*: "I do not define time, space, place and motion, as being well known to all. Only I must observe, that the vulgar conceive those quantities under no other notions but from the relation they bear to sensible objects." In other words, the vulgar among us only understand physical phenomena in terms of folk knowledge. Still, his arguments for gravitation force and planetary motion were constructed in part from aspects of folk physics; specifically, in the *Principia*, Newton relied heavily on spatial and geometric representations, cognitive systems that likely evolved to support large-scale navigation. However, he built on this foundation and corrected naïve explanations through explicit and exacting logic and a period of sustained effort and attention to this work.

The finding that the folk biological classification systems of people living in traditional societies have considerable overlap with the explicit classification systems that emerged with Western science supports the proposal that the former provided the foundation for the latter (Atran, 1998). Yet, here too the relation between folk biology and the biological sciences may be mixed. Inferential biases in folk biology may conspire to make the basic mechanisms of natural selection difficult to comprehend. One bias results in an implicit focus on the behavior of species at different points in a single life span and not the cross-generational time scale over which natural selection occurs. In other words, people may have an inherent bias to think about the biological world in ways that may interfere with comprehending natural selection.

Geometry

The development of geometry—the study of space and shape—as a formal discipline may have been influenced by early geometers' ability to explicitly represent the intuitive knowledge built into folk physical systems that support navigation (Geary, 1995). Euclid (trans. 1956) formally and explicitly postulated a straight line can be draw from any point to any point; that is, the intuitive understanding that the fastest way to get from one place to another is to "go as the crow flies" was made explicit in this postulate. Using a few basic postulates and definitions, Euclid then systematized exciting knowledge to form the often complex and highly spatial components of classic geometry. Euclid's *Elements* must have required an exceptional ability to maintain attentional focus, along with well-developed spatial abilities, and the ability to use logic to explicitly and precisely define spatial relationships.

An important difference between geometry and the physical and biological sciences is that Euclid's (trans. 1956) five postulates and 23 definitions are very basic, defining lines, circles, angles, and shapes, and thus not prone to the naïve attributional errors common in people's descriptions of folk phenomena. These differences may help explain the roughly 1,900- and 2,100-year gaps between Euclid's *Elements* (circa 300 BC) and the 1687 publication of Newton's *Principia* and the 1859 publication of Darwin's *On the Origin of Species*.

Links

My suggestion is the historical emergence of mathematics and the sciences was built from folk domains by individuals with the earlier described combination of traits, such as high intelligence, creativity, ambition, interest in folk physics and biology, and so forth. High intelligence is associated with the ability to explicitly represent and manipulate folk and other information in working memory and to manipulate this information by means of formal logic. When this ability is melded with the scientific method and thus a means to test and correct naïve folk intuitions and attributions, the potential to generate evolutionarily novel domains (e.g., geometry) emerges. We are beginning to understand the brain and cognitive systems that allow people to explicitly represent information in working memory and to systematically manipulate this information in terms of formal rules, and thus build on folk domains (see Geary, 2005).

To illustrate, brain imaging studies suggest areas of the parietal cortex may be one link between evolved spatial abilities and some aspects of mathematical learning, such as forming spatial representations of quantity (Dehaene, Spelke, Pinel, Stanescu, & Tsivkin, 1999). The parietal cortex also appears to be part of the brain systems that support the mental simulation of how objects can be used as tools (Johnson-Frey, 2003). Although a functional relation can only be guessed, it is of interest that areas of parietal cortex typically associated with spatial imagery and other areas of folk physics were unusually large in Einstein's brain (Witelson, Kigar, & Harvey, 1999). In response to a query as to how he approached scientific questions, Einstein replied,

The words of the language . . . do not seem to play any role in my mechanism of thought. The psychical entities which seem to serve as elements in thought are certain signs and more or less clear images which can be "voluntarily" reproduced and combined. . . . There is, of course, a certain connection between those elements and relevant logical concepts. (quoted in Hadamard, 1945, p. 142)

This anecdote is not scientific evidence, of course, but it converges with brain imaging and cognitive research on the relation between folk physical domains and mathematical and scientific reasoning. The implication is that being at the extreme of the folk physical systems that support the use of spatial imagery may contribute to the ability to explicitly represent physical and quantitative information in working memory and may contribute to ease of learning spatial and mechanical aspects of mathematics and the sciences. Indeed, preliminary evidence suggests enhanced folk physical systems may contribute to eminent contributions in these fields (Baron-Cohen, Wheelwright, Stone, & Rutherford, 1999).

Sex Differences

Brain and Cognition

A complete discussion is beyond the scope of this chapter, but consider how an evolutionary perspective can generate hypotheses about sex differences in mathematics and the sciences. One implication of the potential relation between the brain and cognitive systems that support navigation and the ability to use associated spatial representations to solve some forms of mathematical problems is that sex differences in activities that enhance these spatial abilities may incidentally contribute to sex differences in these areas of mathematics. If the evolved activity preferences of boys result in elaboration of these spatial systems during development—and they do (Matthews, 1992)—then boys may have a head start in setting the cognitive foundation for learning in some mathematical and scientific areas.

The relation between activity in areas of the parietal cortex and certain spatial and quantitative abilities may provide a place to begin the systematic testing of this type of hypothesis. Goldstein et al. (2001) found many of these regions are 20% to 25% larger in men than women and have a high density of sex hormone receptors during prenatal development. One implication is that the cognitive and behavioral functions supported by these regions were under stronger selection pressures for men than for women during human evolution, and the organization of this region may be influenced by prenatal exposure to sex hormones. A relation between this region of the parietal cortex and sex differences in geometry and mathematical reasoning has not been established but is an area of potential future study.

Interests and Ambition

As noted, boys are more likely than girls to show an interest in activities that will enhance aspects of folk physical competencies. One hypothesis is these are evolved interest biases that reflect, in part, more frequent tool construction by males than females during human evolution. These biases are predicted to be an aspect of the suite of brain and cognitive systems and developmental activities that allow males to explore and learn about how to construct and use tools from materials in the local ecology. If these sex differences contribute to the sex difference in interest in some areas of the physical sciences and engineering, then within- and between-sex differences in these interests should be predicted by (a) a pattern of interest and engagement in play with mechanical objects during development and (b) prenatal exposure to male hormones.

Moreover, an evolutionary history of more intense intrasexual competition and intersexual vetting in males than females will manifest in modern societies as men being more focused than women on achievement and status in the domains in which they compete (Geary, 1998). In societies with economic specialization, a corollary is that men will define success in terms of areas in which they have a competitive advantage and may accordingly narrow the focus of where they compete. From this perspective, men are also predicted to be more willing than women to trade off other interests, including friends and family, to gain status in these areas; this prediction has been confirmed with Lubinski and Benbow's (chap. 6, this volume) long-term study of mathematically gifted people. The willingness to make these trade-offs, in combination with an interest in mathematics and science domains, are two of the earlier described essential components of exceptional accomplishment.

Within-Sex Variation

On the basis of features of male–male competition, greater within-sex variation is predicted for three-dimensional spatial cognition in men, but on the basis of female–female relational aggression, more variability in language fluency is predicted in women. Most studies have not tested these specific hypotheses but do indicate greater variation on most ability and achievement measures for males than females (Hedges & Nowell, 1995), with the exception of verbal abilities. In a national (U.K.) sample of more than 320,000 children ages 11 to 12 years, Strand, Deary, and Smith (2006) found more boys at the high and low ends of score distributions for measures of quantitative and nonverbal reasoning, but more girls on the high end on a measure of verbal reasoning.

In addition to variation in specific traits, there may be greater variation in many other outcomes, if boys' development and men's functioning are more condition-dependent than those of girls and women. If so, boys and men will be more sensitive to social and environmental conditions for better or worse, which will result in exaggeration of within-sex variation to the extent social and other conditions vary across males. In support, there is evidence that boys' intellectual and academic development and health may be more strongly compromised by poor early environments than those of girls (Martorell, Rivera, Kaplowitz, & Pollitt, 1992). Although we do not know if a sex difference in sensitivity to early conditions contributes to the development of basic abilities (e.g., spatial) and interests that ultimately result in more men than women at the upper end of the distribution for mathematics and science ability, it does provide a framework for systematically approaching these questions.

CONCLUSION

There is now overwhelming evidence that sexual selection (i.e., intrasexual competition and intersexual choice) drives the evolution of most sex differences. These mechanisms result in the exaggeration of traits that facilitate intrasexual competition or influence mate choice and in doing so both create between-sex differences and often increase within-sex variation. The here-and-now expression of these traits results from a combination of exposure to sex hormones, genetic sex, and ecological and social contexts. In fact, some of these traits are condition-dependent; that is, they have evolved to reflect how the individual is coping with ecological and social stressors, and thus their expression is highly dependent on experience and context. For humans, sexual selection is predicted to be influenced by the fact that both sexes often invest in children, which in turn creates intrasexual competition and intersexual choice for both sexes. The devil is in the details though, and human sex differences are predicted to the extent these components of sexual selection have differed for males and females. Any such sex differences are predicted to emerge slowly during childhood and to be influenced by exposure to sex hormones and child-initiated behavioral biases that result in experiences that flesh out these folk competencies and adapt them to local conditions.

Learning the knowledge bases, technical skills, and conceptual models that compose modern mathematics and the sciences is of course more strongly related to schooling and other culture-specific activities than to cognitive evolution per se. However, evolved folk domains and attendant interest biases may provide the foundation for academic learning, and thus our understanding of how people learn mathematics and the sciences, and why they might pursue careers in these fields. Sex differences in folk domains are predicted to the extent these domains are related to parenting, or patterns of intrasexual competition or mate choice. Male–male competition involves greater reliance on the ability to represent three-dimensional space geometrically, and in some mechanical domains related to tool construction. For these areas, average sex differences, favoring boys and men, are predicted, as are more boys and men at the extremes of these abilities and associated interests. Sex differences are predicted for those areas of mathematics and the sciences in which these forms of folk cognition influence learning.

This perspective helps to frame the relation between spatial cognition and mathematical reasoning and the male advantage in both of these areas (Geary, 1996). Any such relation, however, does not lead to a blanket prediction of a male advantage in all areas of mathematics, and in fact provides a means to generate specific and testable hypotheses about where the sexes should be similar and where they should be different. For instance, a male advantage in the ability to visualize mathematical relations in three dimensions is predicted, as is no sex difference in the ability to learn geometric theorems. There are of course many other implications of this perspective that cannot be detailed here. My goal is simply to provide a frame for thinking about the relations between evolved folk knowledge and modern academic learning and for generating hypotheses about when and why sex differences emerge.

REFERENCES

Arnold, A. P., Xu, J., Grisham, W., Chen, X., Kin, Y.-H., & Itoh, Y. (2004). Sex chromosomes and brain sexual differentiation. *Endocrinology, 145*, 1057–1062.

Atran, S. (1998). Folk biology and the anthropology of science: Cognitive universals and cultural particulars. *Behavioral and Brain Sciences, 21*, 547–609.

Baron-Cohen, S., Wheelwright, S., Stone, V., & Rutherford, M. (1999). A mathematician, a physicist and a computer scientist with Asperger syndrome: Performance on folk psychology and folk physics tests. *Neurocase, 5*, 475–483.

Berenbaum, S. A., & Hines, M. (1992). Early androgens are related to childhood sex-typed toy preferences. *Psychological Science, 3*, 203–206.

Bjorklund, D. F., & Pellegrini, A. D. (2002). *The origins of human nature: Evolutionary developmental psychology.* Washington, DC: American Psychological Association.

Blurton Jones, N. G., Hawkes, K., & O'Connell, J. F. (1997). Why do Hadza children forage? In N. L. Segal, G. E. Weisfeld, & C. C. Weisfeld (Eds.), *Uniting psychology and biology: Integrative perspectives on human development* (pp. 279–313). Washington, DC: American Psychological Association.

Chen, Z., & Siegler, R. S. (2000). Across the great divide: Bridging the gap between understanding toddlers' and older children's thinking. *Monographs of the Society for Research in Child Development, 65*(2, Serial No. 261).

Clement, J. (1982). Students' preconceptions in introductory mechanics. *American Journal of Physics, 50*, 66–71.

Clutton-Brock, T. H., & Vincent, A. C. J. (1991, May 2). Sexual selection and the potential reproductive rates of males and females. *Nature, 351*, 58–60.

Cohen-Bendahan, C. C. C., van de Beek, C., & Berenbaum, S. A. (2005). Prenatal sex hormone effects on child and adult sex-typed behavior: Methods and findings. *Neuroscience and Biobehavioral Reviews, 29*, 353–384.

Darwin, C. (1859). *The origin of species by means of natural selection.* London: John Murray.

Darwin, C. (1871). *The descent of man, and selection in relation to sex.* London: John Murray.

Dehaene, S., Spelke, E., Pinel, P., Stanescu, R., & Tsivkin, S. (1999, May 7). Sources of mathematical thinking: Behavioral and brain-imaging evidence. *Science, 284,* 970–974.

Eens, M., & Pinxten, R. (2000). Sex-role reversal in vertebrates: Behavioural and endocrinological accounts. *Behavioural Processes, 51,* 135–147.

Euclid. (1956). *The thirteen books of the elements* (Vol. 1; T. L. Heath, Trans.). New York: Dover. (Original work published circa 300 BC)

Gaulin, S. J. C. (1992). Evolution of sex differences in spatial ability. *Yearbook of Physical Anthropology, 35,* 125–151.

Geary, D. C. (1995). Reflections of evolution and culture in children's cognition: Implications for mathematical development and instruction. *American Psychologist, 50,* 24–37.

Geary, D. C. (1996). Sexual selection and sex differences in mathematical abilities. *Behavioral and Brain Sciences, 19,* 229–284.

Geary, D. C. (1998). *Male, female: The evolution of human sex differences.* Washington, DC: American Psychological Association.

Geary, D. C. (2005). *The origin of mind: Evolution of brain, cognition, and general intelligence.* Washington, DC: American Psychological Association.

Goldstein, J. M., Seidman, L. J., Horton, N. J., Makris, M., Kennedy, D. N., Caviness, V.S., Jr., et al. (2001). Normal sexual dimorphism of the adult human brain assessed by in vivo magnetic resonance imaging. *Cerebral Cortex, 11,* 490–497.

Golombok, S., & Rust, J. (1993). The pre-school activities inventory: A standardized assessment of gender role in children. *Psychological Assessment, 5,* 131–136.

Gredlein, J. M., & Bjorklund, D. F. (2005). Sex differences in young children's use of tools in a problem-solving task. *Human Nature, 16,* 211–232.

Hadamard, J. (1945). *The psychology of invention in the mathematical field.* New York: Dover.

Hedges, L. V., & Nowell, A. (1995, July 7). Sex differences in mental scores, variability, and numbers of high-scoring individuals. *Science, 269,* 41–45.

Hill, S. T., & Johnson, J. M. (2004, March). *Science and engineering degrees: 1966–2001.* Arlington, VA: National Science Foundation. Retrieved July 13, 2006, from http://www.nsf.gov/statistics/nsf04311/

Hines, M., Fane, B. A., Pasterski, V. L., Mathews, G. A., Conway, G. S., & Brook, C. (2003). Spatial abilities following prenatal androgen abnormality: Targeting and mental rotations performance in individuals with congenital adrenal hyperplasia. *Psychoneuroendocrinology, 28,* 1010–1026.

Hines, M., Golombok, S., Rust, J., Johnston, K. J., Golding, J., & the Avon Longitudinal Study of Parents and Children Team. (2002). Testosterone during pregnancy and gender role behavior in preschool children: A longitudinal, population study. *Child Development, 73,* 1678–1687.

Hirschfeld, L. A., & Gelman, S. A. (Eds.). (1994). *Mapping the mind: Domain specificity in cognition and culture.* New York: Cambridge University Press.

Iervolino, A. C., Hines, M., Golombok, S. E., Rust, J., & Plomin, R. (2005). Genetic and environmental influences on sex-typed behavior during the preschool years. *Child Development, 76,* 826–840.

Johnson-Frey, S. H. (2003). What's so special about human tool use? *Neuron, 39,* 201–204.

Keeley, L. H. (1996). *War before civilization: The myth of the peaceful savage.* New York: Oxford University Press.

Martorell, R., Rivera, J., Kaplowitz, H., & Pollitt, E. (1992). Long-term consequences of growth retardation during early childhood. In M. Hernández & J. Argente (Eds.), *Human growth: Basic and clinical aspects* (pp. 143–149). Amsterdam: Elsevier Science.

Matthews, M. H. (1992). *Making sense of place: Children's understanding of large-scale environments.* New York: Rowman & Littlefield.

Morris, J. A., Jordan, C. L., & Breedlove, S. M. (2004). Sexual differentiation of the vertebrate nervous system. *Nature Neuroscience, 7,* 1034–1039.

Murdock, G. P. (1981). *Atlas of world cultures.* Pittsburgh, PA: University of Pittsburgh Press.

Murray, C. (2003). *Human accomplishment: The pursuit of excellence in the arts and sciences, 800 B.C. to 1950.* New York: HarperCollins.

Newton, I. (1995). *The principia* (A. Motte, Trans.). Amherst, NY: Prometheus Books. (Original work published 1687)

Nuttall, R. L., Casey, M. B., & Pezaris, E. (2005). Spatial ability as a mediator of gender differences on mathematics tests. In A. M. Gallagher & J. C. Kaufman (Eds.), *Gender differences in mathematics: An integrative psychological approach* (pp. 121–142). Cambridge, England: Cambridge University Press.

Pinker, S. (1997). *How the mind works.* New York: Norton.

Pomiankowski, A., & Møller, A. P. (1995). A resolution of the lek paradox. *Proceedings of the Royal Society of London B, 260,* 21–29.

Resnick, S. M., Berenbaum, S. A., Gottesman, I. I., & Bouchard, T. J., Jr. (1986). Early hormonal influences on cognitive functioning in congenital adrenal hyperplasia. *Developmental Psychology, 22,* 191–198.

Shepard, R. N. (1994). Perceptual–cognitive universals as reflections of the world. *Psychonomic Bulletin & Review, 1,* 2–28.

Simonton, D. K. (1999). *Origins of genius: Darwinian perspective on creativity.* New York: Oxford University Press.

Spelke, E. S. (2005). Sex differences in intrinsic aptitude for mathematics and science: A critical review. *American Psychologist, 60,* 950–958.

Strand, S., Deary, I. J., & Smith, P. (2006). Sex differences in cognitive abilities test scores: A UK national picture. *British Journal of Educational Psychology, 76*, 463–480.

Stumpf, H., & Stanley, J. C. (1998). Stability and change in gender-related differences on the college board advanced placement and achievement tests. *Current Directions in Psychological Science, 7*, 192–196.

Trivers, R. L. (1972). Parental investment and sexual selection. In B. Campbell (Ed.), *Sexual selection and the descent of man 1871–1971* (pp. 136–179). Chicago: Aldine.

Whiting, B. B., & Edwards, C. P. (1988). *Children of different worlds: The formation of social behavior.* Cambridge, MA: Harvard University Press.

Willingham, W. W., & Cole, N. S. (1997). *Gender and fair assessment.* Mahwah, NJ: Erlbaum.

Witelson, S. F., Kigar, D. L., & Harvey, T. (1999). The exceptional brain of Albert Einstein. *Lancet, 353*, 2149–2153.

14

NEURAL SUBSTRATES FOR SEX DIFFERENCES IN COGNITION

RUBEN C. GUR AND RAQUEL E. GUR

Efforts to understand the biological substrates of sex differences in cognition require examination of brain anatomy (structure) and physiology (function). Technological and methodological advances have increasingly enabled the examination in humans of the neurobiology of behavior across the life span. Starting in the early 1980s, a genre of safe methods for obtaining reliable measures of brain structure and function have become available. Although most applications of these methods have been in people with different brain disorders, several sufficiently large-scale efforts have included healthy people and have examined sex differences in brain anatomy and physiology. Fewer studies have related such measures to cognitive performance; nonetheless, there is considerable convergence of replicated findings to support at least some hypotheses worthy of further refinement. A review of this literature is beyond the scope of this chapter; instead we briefly describe the main findings from neuroimaging applications in which sex differences were established. These findings substantiate some hypotheses on neural substrates for sex differences in cognition.

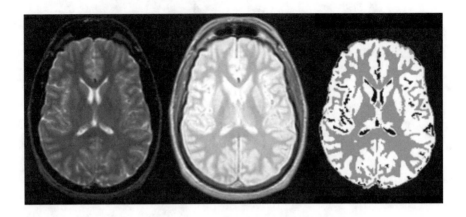

Figure 14.1. Illustration of the magnetic resonance imaging segmentation process showing an acquired T2-weighted image (left), a proton density image (middle), and the segmented image (right) in which gray matter is depicted in white, white matter in light gray, and cerebrospinal fluid in black. From "Sex Differences in Brain Gray and White Matter in Healthy Young Adults: Correlations With Cognitive Performance," by R. C. Gur et al., 1999, *Journal of Neuroscience, 19,* p. 4066. Copyright 1999 by the Society for Neuroscience. Reprinted with permission.

BRAIN ANATOMY

Sex Differences in Cranial Tissue Composition

One can begin the search for anatomic differences by measuring the tissue in our brain, which is composed of gray matter (GM; the somatodendritic tissue of neurons, namely the cell body and the many short protrusions, called *dendrites,* through which the nerve cell receives input from adjacent cells) and white matter (WM; which is composed of connecting fibers through which neurons communicate and that are wrapped in myelin, a fatty tissue). The brain is surrounded by cerebrospinal fluid (CSF), the slow circulating fluid that surrounds both cortical and spinal neuronal tissue. Initial studies that suggested sex differences in the composition of human neural tissue used a noninvasive procedure that can measure the proportion of tissue with fast blood flow (presumably GM). These studies showed rather substantial sex differences in the percentage of GM in the cortical surface, with women having higher values (R. C. Gur et al., 1982). The main current method for studying brain anatomy is magnetic resonance imaging (MRI). Structural MRI studies use a variety of methods for segmentation of tissue into GM, both cortical and deep, WM, and CSF (Figure 14.1). Replicating the earlier findings, higher percentage of GM was found in women, but with MRI it was also possible to establish that men had higher percentages of WM and CSF (Coffey et al., 1998; R. C. Gur et al., 1999; Figure 14.2, top panels).

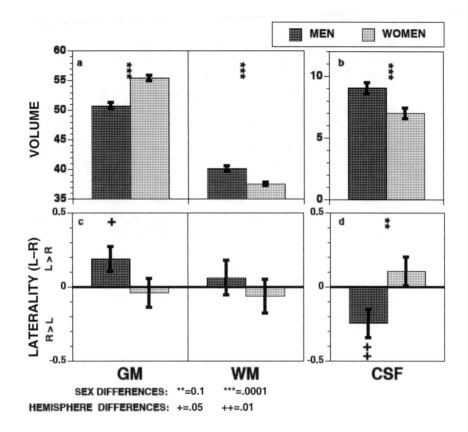

Figure 14.2. Means ± *SEM* percentage of tissue and cerebrospinal fluid (CSF) averaged bilaterally (top) and examined as a laterality index (left [L] minus right [R], bottom) in men (dark bars) and women (light bars). Women have a higher percentage of gray matter (GM), whereas men have a higher percentage of white matter (WM) and CSF. Furthermore, men have a significantly higher percentage of GM on the left and CSF on the right, whereas in women all brain compartments are symmetric. From "Sex Differences in Brain Gray and White Matter in Healthy Young Adults: Correlations With Cognitive Performance," by R. C. Gur et al., 1999, *Journal of Neuroscience, 19,* p. 4068. Copyright 1999 by the Society for Neuroscience. Reprinted with permission.

Sex differences in hemispheric asymmetries were also documented, with greater asymmetries in the percentage of GM and CSF in men compared with women (Coffey et al., 1998; R. C. Gur et al., 1999; Figure 14.2, bottom panels). For men, the percentage of GM was higher in the left, WM was symmetric, but the percentage of CSF was higher on the right. No asymmetries were significant in women, and the difference in *laterality gradients* (the difference in volume between two hemispheres) between men and women was significant. The hemispheric effects were quite small in absolute terms and did not overshadow the main sex differences in raw volumes. Thus, whereas men had higher percentage GM in the left relative to the right hemisphere and women had symmetric GM, women still had a higher percentage of GM than men in either hemisphere. Because GM can be considered as the hardware necessary for computation whereas WM serves for communicating among computational nodes, it can be argued that women have compensated for smaller cranial volume by packing a higher percentage of tissue with computational power.

Few studies have examined sex differences in the correlation between cognitive performance and the volume of intracranial compartments. Anatomic findings may provide neural substrates for sex differences in cognition if volume correlates with performance on cognitive tasks. We examined whether our sample (R. C. Gur et al., 1999) showed the reported sex difference of better verbal relative to spatial performance in women compared with men. Men and women did not differ in the global performance score. However, as expected, the verbal superiority index (verbal minus spatial) was positive in women and negative in men, and the two groups differed. Further supporting the functional significance of the neuroanatomic findings, cognitive performance correlated with intracranial volumes for the whole sample and similarly for men and women.

Corpus Callosum

The corpus callosum is a bundle of nerve fibers that connects the two brain hemispheres and is necessary for interhemispheric communication. In contrast to globally lower WM volume in women, there is some evidence that the corpus callosum, which is the largest WM structure in the brain, is larger or at least more bulbous in women. Cognitive and functional imaging studies have suggested a greater degree of hemispheric lateralization in males compared with females, whereas females displayed increased bilateral hemispheric activity for a variety of cognitive tasks (reviewed in Kimura & Harshman, 1984). These studies seem to suggest enhanced interhemispheric communication in females and have motivated investigation into sexual dimorphism of the corpus callosum. Most investigators have examined the shape and size of the midsagittal section of the callosum as a surrogate for the structure's overall shape. To date, no consensus has been reached on the presence of such gender-based differences in the callosum. A possible reason for this controversy is the lack of standards in callosal analysis. Template deformation morphometry (TDM) is a relatively new method that avoids many of the pitfalls associated with more traditional measures of the callosum. By registering each subject to a template callosum, TDM avoids the issue of normalizing callosal measurements to some arbitrary index of overall brain size. TDM demonstrated that the splenium (back section) of the callosum was larger in females than males, whereas a relatively larger genu (the middle section) of the callosum was found in males (Figure 14.3).

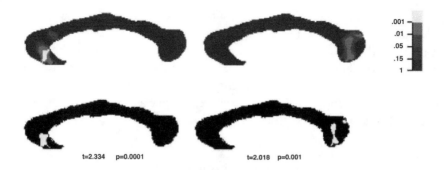

Figure 14.3. Size comparison of male and female callosa. Top row displays raw pointwise P values, and bottom row displays corrected clusters with P values. Left side shows areas of larger male size and right shows areas of larger female size. From "Characterization of Sexual Dimorphism in the Human Corpus Callosum," by A. Dubb, R. C. Gur, B. Avants, and J. Gee, 2003, *Neuroimage, 20,* p. 515. Copyright 2003 by Elsevier. Reprinted with permission.

The finding of the larger female splenium is consistent with the arrangement of the corpus callosum in which the front connects frontal brain regions and the back connects posterior visual cortex regions. The neuropsychological literature indicates enhanced bihemispheric representation of verbal tasks in females (e.g., R. C. Gur et al., 2000; Shaywitz et al., 1995), and the splenium would be involved in interhemispheric transfer of language processing. The finding of the larger male genu may relate to enhanced motor coordination in men, a finding also supported by the literature (Maccoby, 1966).

Longitudinal studies and studies of young infants (Matsuzawa et al., 2001) may help elucidate causal direction of the relation between brain volume and cognition. Giedd et al. (1999) demonstrated sex differences in the pattern of GM and WM development. The curves were similar but tended to peak at different ages. The peaks tended to be earlier (e.g., in terms of peak GM volume for frontal cortex) for girls than boys, except for the temporal cortex in which girls peaked at a slightly older age. A notable pattern was that occipital GM had not yet peaked for males by age 22, but had peaked at about age 13 for females. WM increased for both sexes from 4 to 22 years but at a higher rate for males than females. The increased brain-development period for males, especially with respect to WM and occipital GM, is intriguing because these correlate with spatial performance in adults. The extended developmental period also makes male brain development more condition dependent—good health is needed for a longer period of time to achieve full potential.

To summarize the anatomic studies, sex differences are evident across the age range, and there are sex differences in age effects on brain compartmental volumes. In general, women have a higher percentage of tissue devoted to neuronal cell bodies and their immediate dendritic connections, whereas men have a higher volume of connecting WM tissue, with the exception of the splenium of the corpus callosum, which is more bulbous in women than men. Furthermore, male brains show greater volumetric asymmetries than female brains. The higher WM volume seems associated with better spatial performance in men, whereas lower asymmetry seems associated with better language processing in women.

BRAIN PHYSIOLOGY

The feasibility of studying neural substrates of behavior is enhanced by functional neuroimaging methods for measuring regional brain activity. Activation patterns are linked to performance on cognitive tasks requiring cognitive operations such as verbal, spatial, attention, memory, and facial processing. Performance measures are obtained during the physiological studies, permitting linkage of brain activity with task execution.

Sex Differences in Cerebral Blood Flow

Sex differences have not been examined as extensively with functional as with structural imaging. Consistently across studies, women have higher rates of resting cerebral blood flow (CBF) than men (R. C. Gur et al., 1982). Sex differences in activation patterns are less consistent (for discussions, see R. E. Gur & Gur, 1990; Kastrup, Li, Glover, Kruger, & Moseley, 1999). Greater bilateral activation for language tasks was reported in females (Shaywitz et al., 1995). For spatial tasks, better performance of men when solving the harder problems was associated with more focal activation of right visual association areas (R. C. Gur et al., 2000). In contrast, women recruited additional regions bilaterally for the harder spatial task. This finding was replicated and extended to mental rotation and numeric calculation tasks by Kucian, Loenneker, Dietrich, Martin, and von Aster (2005), who likewise reported more distributed and bilateral recruitment of regions in women than men with increased task complexity. Similarly, Grön, Wunderlich, Spitzer, Tomczak, and Riepe (2000) reported that men and women used different brain regions in a three-dimensional virtual maze. Women used more parietal and prefrontal regions (the latter suggesting it was an effortful task), whereas the men relied more on the hippocampus, suggesting an automatic encoding of geometric-navigation cues.

Sex Differences in Cerebral Glucose Metabolism

In contrast to higher global CBF in women, resting cerebral metabolic rates for glucose (CMRglu) are equal in men and women (Andreason,

Zametkin, Guo, Baldwin, & Cohen, 1994; R. C. Gur et al., 1995; Murphy et al., 1996). Sex differences are evident in the regional distribution of metabolic activity, with men showing higher glucose metabolism in all basal ganglia regions, which are subcortical nuclei that participate in the control of movement, and the cerebellum, which consists of the hindbrain also responsible for movement, muscle tone, and balance. Men also had higher metabolic activity in all limbic regions that are below the corpus callosum, whereas women showed higher metabolism in the cingulate gyrus, a more phylogenetically advanced limbic region closer to language areas. As with MRI, women show more symmetric glucose utilization than men (R. C. Gur et al., 1995; Murphy et al., 1996).

Sex Differences in Neurotransmitter Function

Another set of physiologic parameters that can be measured with functional neuroimaging is neurotransmitter function. Depending on the specific neurotransmitter, greater abundance or receptor availability can facilitate or inhibit brain function. Few studies included sufficiently large samples to examine sex differences. Of these, Adams et al. (2004) reported no sex differences in serotonin (5-HT) binding. However, sex differences were found in dopamine function. The dopamine transporter is the primary indicator of dopaminergic tone, and Mozley, Gur, Mozley, and Gur (2001) investigated the relationship between cognition and dopamine transporter availability in healthy men and women. Women had higher dopamine availability in the caudate nucleus, and they also performed better on verbal learning tasks. Furthermore, dopamine transporter availability was correlated with learning performance within groups.

To summarize the studies on sex differences in brain physiology, men and women differ both in the resting state and in the topography of activation in response to tasks. At a resting state, women have higher rates of blood flow but similar rates of glucose metabolism. The increased blood flow relative to metabolism may help protect the brain from episodes of insufficient oxygen supply that can damage brain cells. During task performance, women tend to show more bilateral activation, consistent with the anatomic findings of diminished asymmetry. This bilateral activation may confer advantages in tasks that require bihemispheric participation, which seems to be the case for aspects of language processing. Women also appear to have greater availability of dopamine transporters, which relates to better memory.

HYPOTHESES ON BEHAVIORAL EFFECTS OF SEX DIFFERENCES IN BRAIN ANATOMY AND PHYSIOLOGY

The state of knowledge on the neurobiology of sex differences is far from enabling strong statements. Especially lacking are large-scale studies in healthy people in which behavioral data are rigorously measured and related to brain anatomy and physiology. Nonetheless, several tentative hypotheses can be proposed.

A hypothesis based on the neuroanatomic data is that male brains are optimized for enhanced connectivity within hemispheres presumably in the anterior–posterior or dorsal–ventral directions, as afforded by overall higher WM volumes, whereas female brains are optimized for communication between the hemispheres, especially in language processing and posterior brain regions, as indicated by the larger callosal splenia. Evolutionarily, this may have conferred males advantage in actions requiring rapid transition from perception (posterior) to action (anterior), incorporating limbic (ventral) input. For females, better interhemispheric communication confers advantage in language and the ability to better integrate verbal–analytical (left-hemispheric) with spatial–holistic (right-hemispheric) modes of information processing. This hypothesis can be tested directly with novel MRI methods using diffusion tensor imaging.

Biologically, females have higher CBF and the same metabolic rates as males. This affords them "luxury perfusion" relative to metabolic demands and may better equip them for sustained mental activity. The excess CBF relative to CMRglu may also confer longevity of tissue, which may relate to the higher longevity of women. Activation studies support the notion that women perform better on tasks requiring bilateral activation, such as language processing, whereas men excel in tasks requiring focal activation of visual association cortex.

Regarding the question of the differential representation of women in scientific, technical, engineering, and math (STEM) careers, biology can only offer a limited perspective and at this point not much more than conjectures. The requirement of large volume of WM for complex spatial processing may be an obstacle in some branches of mathematics and physics. However, the greater facility of women with interhemispheric communications may attract them to disciplines that require integration rather than detailed scrutiny of narrowly characterized processes. These are highly valued in STEM fields. Notably, however, there are large branches of mathematics that do not require spatial processing, and perhaps as more women enter the field, these branches will receive more attention and may turn out to be of value comparable with spatial domains. The biological findings may also explain some of the motivational factors underlying lower representation of women in STEM careers. Increased activity in limbic regions related to aggression in men may confer an advantage in highly competitive fields. One should bear in mind that biology reflects an evolutionary epoch much larger than the brief time during which women were allowed to compete freely in the intellectual marketplace, and that attitudes discouraging participation of women are stronger in STEM careers than in the social sciences and humanities. We need many more years of a level playing field to rule out such attitudes as accounting for a much larger portion of the variance than biology will ever explain.

REFERENCES

Adams, K. H., Pinborg, L. H., Svarer, C., Hasselbalch, S. G., Holm, S., Haugbol, S., et al. (2004). A database of [(18)F]-altanserin binding to 5-HT(2A) receptors in normal volunteers: Normative data and relationship to physiological and demographic variables. *NeuroImage, 21,* 1105–1113.

Andreason, P. J., Zametkin, A. J., Guo, A. C., Baldwin, P., & Cohen, R. M. (1994). Gender-related differences in regional cerebral glucose metabolism in normal volunteers. *Psychiatry Research, 51,* 175–183.

Coffey, C. E., Lucke, J. F., Saxton, J. A., Ratcliff, G., Unitas, L. J., Billig, B., et al. (1998). Sex differences in brain aging: A quantitative magnetic resonance imaging study. *Archives of Neurology, 55,* 169–179.

Dubb, A., Gur, R. C., Avants, B., & Gee, J. (2003). Characterization of sexual dimorphism in the human corpus callosum. *NeuroImage, 20,* 512–519.

Giedd, J. N., Blumenthal, J., Jeffries, N. O., Castellanos, F. X., Liu, H., Zijdenbos, A., et al. (1999). Brain development during childhood and adolescence: A longitudinal MRI study. *Nature Neuroscience, 2,* 861–863.

Grön, G., Wunderlich, A. P., Spitzer, M., Tomczak, R., & Riepe, M. W. (2000). Brain activation during human navigation: Gender-different neural networks as substrate of performance. *Nature Neuroscience, 3,* 404–408.

Gur, R. C., Alsop, D., Glahn, D., Petty, R., Swanson, C. L., Maldjian, J. A., et al. (2000). An fMRI study of sex differences in regional activation to a verbal and a spatial task. *Brain and Language, 74,* 157–170.

Gur, R. C., Gur, R. E., Obrist, W. D., Hungerbuhler, J. P., Younkin, D., Rosen, A. D., et al. (1982, August 13). Sex and handedness differences in cerebral blood flow during rest and cognitive activity. *Science, 217,* 659–661.

Gur, R. C., Mozley L. H., Mozley, P. D., Resnick, S. M., Karp, J. S., Alavi, A., et al. (1995, January 27). Sex differences in regional cerebral glucose metabolism during a resting state. *Science, 267,* 528–531.

Gur, R. C., Turetsky, B. I., Matsui, M., Yan, M., Bilker, W., Hughett, P., et al. (1999). Sex differences in brain gray and white matter in healthy young adults: Correlations with cognitive performance. *Journal of Neuroscience, 19,* 4065–4072.

Gur, R. E., & Gur, R. C. (1990). Gender differences in regional cerebral blood flow. *Schizophrenia Bulletin, 16,* 247–254.

Kastrup, A., Li, T. Q., Glover, G. H., Kruger, G., & Moseley, M. E. (1999). Gender differences in cerebral blood flow and oxygenation response during focal physiologic neural activity. *Journal of Cerebral Blood Flow Metabolism, 19,* 1066–1071.

Kimura, D., & Harshman, R. A. (1984). Sex differences in brain organization for verbal and non-verbal functions. *Progress in Brain Research, 61,* 423–441.

Kucian, K., Loenneker, T., Dietrich, T., Martin, E., & von Aster, M. (2005). Gender differences in brain activation patterns during mental rotation and number related cognitive tasks. *Psychology Science, 47,* 112–131.

Maccoby, E. (1966). *Development of sex differences.* Stanford, CA: Stanford University Press.

Matsuzawa, M., Matsui, M., Konishi, T., Noguchi, N., Gur, R. C., Bilker, W., et al. (2001). Age-related volumetric changes of brain gray and white matter in healthy infants and children. *Cerebral Cortex, 11,* 335–342.

Mozley, L. H., Gur, R. C., Mozley, P. D., & Gur, R. E. (2001). Striatal dopamine transporters and cognitive functioning in healthy men and women. *American Journal of Psychiatry, 158,* 1492–1499.

Murphy, D. G., DeCarli, C., McIntosh, A. R., Daly, E., Mentis, M. J., Pietrini, P., et al. (1996). Sex differences in human brain morphometry and metabolism: An in vivo quantitative magnetic resonance imaging and positron emission tomography study on the effect of aging. *Archives of General Psychiatry, 53,* 585–594.

Shaywitz, B. A., Shaywitz, S. E., Pugh, K. R., Constable, R. T., Skudlarski, P., Fulbright, R. K., et al. (1995, February 16). Sex differences in the functional organization of the brain for language. *Nature, 373,* 607–609.

15

WHERE ARE ALL THE WOMEN? GENDER DIFFERENCES IN PARTICIPATION IN PHYSICAL SCIENCE AND ENGINEERING

JACQUELYNNE S. ECCLES

The president of Harvard University, Lawrence Summers, gave a speech in 2005 that set off a firestorm of discussion about the relatively low numbers of women researchers in the fields of science, engineering, and math. I found this debate particularly interesting because I have been studying exactly this issue for the past 35 years. I began my research career in the late 1970s with a grant from the then National Institute of Education specifically focused on the relatively low numbers of women at the most advanced levels of mathematics-related professions. A great deal of research was done on this topic at that time, and considerable funds were put into designing and testing a wide variety of intervention programs to increase female participation in mathematics-related careers, such as engineering and physical science. Great progress was made, moving the average enrollment of women in bachelor's degree programs in engineering from a little more than 3% to approximately 18% by the 1990s. It is interesting, both then and now, that women have always been well represented in mathematics itself, which turns out to be one of the least sex-typed undergraduate college majors. Furthermore, for at least the past 10 years, women have been very well represented in both undergraduate- and master's-level programs in the biological and medical sciences, as well as in MD programs. Nevertheless, women continue to be underrepresented in programs in physical science and engineering and on the university and college faculties in all of the natural sciences, engineering, and mathematics. Why?

GENERAL EXPECTANCY VALUE MODEL

As noted earlier, my colleagues and I have studied the psychological and social factors influencing course enrollment decisions, college major selection, career aspirations, and career choices for the past 35 years. I began this work with a particular interest in the psychological and social factors that might underlie the gender differences in educational and vocational choices, particularly in the fields of mathematics, physical science, and engineering. Frustrated with the many disconnected theories emerging to explain such gender differences, my colleagues and I developed a comprehensive theoretical model of achievement-related choices (Eccles Parsons et al., 1983; see Figure 15.1 for most recent version). Drawing on work associated with decision making, achievement theory, and attribution theory, we hypothesized that the kinds of educational and vocational decisions that might underlie gender differences in participation in physical science and engineering would be most directly influenced by individuals' expectations for success and the importance or value individuals attach to the various options they see as available. We then hypothesized how these quite domain-specific self- and task-related beliefs might be influenced by cultural norms, experiences, aptitudes, and more general personal beliefs (see Eccles, 1994; Eccles Parsons et al., 1983; Eccles, Wigfield, & Schiefele, 1998). We have used this model as a guide to our program of research ever since.

For example, consider decisions related to selecting a college major. According to our model, people should be most likely to choose a major that they think they can master and that has high task value for them. Expectations for success (domain-specific beliefs about one's personal efficacy to master the task), in turn, depend on both the confidence that individuals have in their various intellectual abilities and the individuals' estimations of the difficulty of the various options they are considering. We also predicted that these self- and task-related beliefs were shaped over time by both experiences with the related school subjects and activities and individuals' subjective interpretation of these experiences (e.g., does the person think that her or his prior successes reflect high ability or lots of hard work? And if the latter, will it take even more work to continue to be successful?). As I discuss in more detail later, gender role socialization and commonly held stereotypes

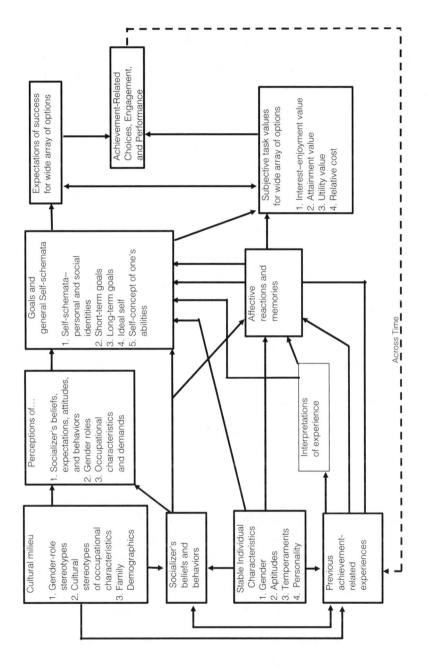

Figure 15.1. Eccles expectancy value model of achievement-related choices. Copyright 1983 by J. Eccles (Parsons). Printed with permission.

about gender differences in "natural talents" for various subject areas are likely to lead females and males to have different estimates of their own personal efficacies for physical science and engineering. It is quite likely that females will receive less support for developing a strong sense of their talent for these fields from their parents, teachers, and peers than males.

Likewise, our model specifies that the subjective task value of various majors is influenced by several factors. For example, does the person enjoy doing the related school work? Is this major seen as instrumental in meeting the individual's long- or short-range goals? Have the individual's parents, counselors, friends, or romantic partners encouraged or discouraged the individual from selecting this major? Does taking the major interfere with other more valued options because of the amount of work needed to be successful either in the major or in the future professions linked to the major? Again, as discussed in more detail later, it is quite likely that males will receive more support for developing a strong interest in physical science and engineering from their parents, teachers, and peers than females. In addition, it is absolutely the case that all young people will see more examples of males engaged in these occupations than females. Consequently, according to our model, the likelihood of even considering these occupations as appropriate is much lower for females than for males.

SUBJECTIVE TASK VALUE

As we developed our model in the mid-1970s, it became clear to us that the theoretical grounding for understanding the nature of subjective task value was much less well developed than the theoretical grounding for understanding the nature of expectations for success. Consequently, we elaborated our notion of subjective task value to help fill this void. Drawing on work associated with achievement motivation, intrinsic versus extrinsic motivation, self psychology, identity formation, economics, and organization psychology, we hypothesized that subjective task value was composed of at least four components: interest value (the enjoyment one gets from engaging in the task or activity), utility value (the instrumental value of the task or activity for helping to fulfill another short- or long-range goal), attainment value (the link between the task and one's sense of self and identity), and cost (defined in terms of either what may be given up by making a specific choice or the negative experiences associated with a particular choice).

Because I believe that the last three of these are particularly relevant for understanding gender differences in educational and occupational choices, I elaborate on these further now. My colleagues and I have argued that the socialization processes linked to gender roles are likely to influence both short- and long-term goals and the characteristics and values most closely linked to core identities (e.g., Eccles, 1992, 1994; Jacobs & Eccles, 1992). For example, gender role socialization is likely to lead to gender differences in the kinds of work one would like to do as an adult: Females should be more likely than males to want to work at occupations that help others and fit well into their family role plans. Males should be more likely than females to want future occupations that pay very well and provide opportunities to become famous (for reviews of evidence supporting these hypotheses, see Eccles et al., 1998; Ruble & Martin, 1998; for empirical support, see Eccles & Vida, 2003). There is also evidence that males are somewhat more interested than females in activities and jobs related to manipulating physical objects and abstract concepts, whereas females are more interested in activities and jobs related to people and social interactions. These differences are likely to influence the types of jobs that appeal to male and female adolescents as they are in the process of making related educational decisions. If so, then the utility value and attainment value of various high school courses and college majors should differ on average by gender, precisely because these courses and majors are linked directly to adult occupational choices.

Similarly, the perceived cost of different high school courses and college majors should vary by gender because of the relative importance attached to various options. The cost may also vary by gender because of average-level differences in such emotional costs as math anxiety and the fear of rejection for making nontraditional choices (see Eccles et al., 1998).

Three features of our approach that are not well captured by the static model depicted in Figure 15.1 are particularly important for understanding gender differences in the types of educational and occupational choices represented in this book: First, we are interested in both conscious and nonconscious achievement-related behavioral choices. Although the language we use to describe the various components makes it seem that we are talking about quite conscious processes, this is not our intention. We believe that the conscious and nonconscious choices people make about how to spend their time and effort lead, over time, to marked differences between groups and individuals in lifelong achievement-related patterns. We also believe, however, that these choices are heavily influenced by socialization pressures and cultural norms.

Second, we are interested in what becomes part of an individual's perception of viable options. Although individuals choose from among several options, they do not actively consider the full range of objectively available options. Many options may never be considered because the individual is either unaware of their existence or has little opportunity or encouragement to really think about a wide range of alternatives. Other options may not be seriously considered because the individual has inaccurate information regarding either the option itself or the individual's potential for achieving the option. For example, young people often have inaccurate information regarding the full range of activities associated with various career choices or inaccurate information regarding the financial assistance available for advanced educational training. Yet they make decisions about which occupations to pursue and then select courses in high school that they believe are important for getting into college and majoring in the subject most directly linked to their career aspirations. Too often these choices are based on either inaccurate or insufficient information. Finally, many options may not be seriously considered because the individual does not believe that a particular choice fits well with his or her gender role or other social role schemas. Again, inaccurate information about what occupations are actually like can lead to premature elimination of quite viable career options. For example, a young woman with excellent math skills may reject the possibility of becoming an engineer or a computer scientist because she has a limited view of what engineers and computer scientists actually do. She may stereotype engineers as nerds or as people who focus on mechanical tasks with little direct human relevance, when in fact many engineers work directly on problems related to pressing human needs. If so, she may well select herself, or be encouraged to select herself, out of a profession that she might both enjoy and find quite compatible with her life goals and values. As a culture, we do a very poor job of providing information to our children and adolescents about various occupations. As a consequence, they must rely on media portrayals and happenstance career counseling from their parents, mentors, and friends. Such portrayals are often quite gender and ethnically stereotyped.

Third, we assume that educational and occupational decisions are made within a complex social reality. For example, the decision to major in biology rather than computer science or engineering is made within the context of a complex social reality that presents each individual with a wide variety of choices, each of which has both immediate and long-range consequences that map in complex ways onto the full range of determinants of subjective task value. Furthermore, many options have both positive and negative components. For example, the decision to enroll in an advanced math course in high school is typically made in the context of other important daily-life decisions and long-term life decisions such as whether to take advanced English to study literature one enjoys, to take a second foreign language course to aid in one's future travel plans, to take a course with one's best friend or romantic partner to have an intellectual activity to share, or to take less demanding courses to spend more time enjoying the social aspects of one's senior year. Similarly, the decision to major in computer science or engineering versus something else is made in the context of a wide variety of options and life demands during the college years. The critical issues in our view are the relative personal value of each option and the individual's assessment of his or her relative abilities and potentials at the time the decision is being made. In addition, having narrowed the field to those options at which one feels confident about succeeding, we assume that people will then choose the options with the highest personal value. Thus, it is the hierarchy of subjective task values and expectations for success that matter rather than the absolute values of both of these belief systems that are attached to the various options under consideration. This feature of our approach makes within-person comparisons much more relevant to understanding individuals' decisions than between-group mean-level comparisons. Unfortunately, very little work has taken such a pattern-centered approach.

Consider two high school students, Mary and Beth. Both young women enjoy mathematics and physical science and have always done very well in these subjects, as well as in their other school subjects. Both have been identified as gifted in mathematics and have been offered the opportunity to participate in an accelerated math program at the local college during their senior year. Beth hopes to major in communications when she gets to college and has also been offered the opportunity to work part time at the local TV news station doing odd jobs and some copy editing. Mary hopes to major in computer science in college and plans a career as a research scientist designing educational software. Taking the accelerated math course involves driving to and from the college. Because the course is scheduled for the last period of the day, it will take the last two periods of the day as well as 1 hour of after-school time to take the course. What will the young women do? In all likelihood, Mary will enroll in the program because she both likes math and thinks that the effort required to master the material is important for her long-range career goals. Beth's decision is more complex. She may want to take the class but may also think that the time required is too costly, especially given her alternative opportunity to do an apprenticeship at the local TV station. Whether she takes the college course or not will depend, in part, on the advice she gets from her counselors, family, and friends. If they stress the importance of the math course, then its subjective worth is likely to increase. If the subjective worth of the course increases sufficiently to outweigh its subjective cost, then Beth will likely take the course despite its cost in time and effort. Studying these types of subtle processes is difficult with individual- and group-difference-oriented, variable-centered approaches.

In summary, my colleagues and I assume that educational and occupational choices (as well as other achievement-related leisure-time choices) are guided by (a) individuals' expectations for success on (sense of personal efficacy at) the various options, as well as their sense of competence for various tasks; (b) the relation of the options to their short- and long-range goals, to their core personal and social identities, and to their basic psychological needs; (c) individuals' culturally based role schemas linked to gender, social class, and ethnic group; and (d) the potential cost of investing time in one activity rather than another. We assume that all of these psychological variables are influenced by individuals' histories as mediated by their interpretation of these experiences, by cultural norms, and by the behaviors and goals of one's socializers and peers.

MICHIGAN STUDY OF ADOLESCENT LIFE TRANSITIONS: A STUDY OF THE ECCLES EXPECTANCY-VALUE MODEL OF ACHIEVEMENT TASK CHOICES

We have spent the past 35 years amassing evidence to support each hypothesis. The findings related to gender differences in the pursuit of careers in physical science and engineering are quite robust (see Eccles, 1992, 1994; Eccles et al., 1998, for reviews). Here I give just one example of our most recent findings. The analyses I report were done using data from the Michigan Study of Adolescent Life Transitions. This longitudinal study is being conducted by myself and Bonnie Barber. It began in 1982 with a sample of approximately 3,000 sixth graders in 12 different school districts in southeastern Michigan; these districts served primarily working-class and middle-class small-city communities. The sample is predominantly White but does include about 150 African American adolescents. We have now followed approximately 1,500 of these adolescents well into their early adulthood years, using standard survey type methods. All of the survey instruments reported here have been used in a variety of studies and have well-established reliability and good predictive and face validity.

First, we looked at the psychological predictors of enrollment in both the honor's mathematics track in high school and high school physics. We found no gender differences in enrollment in high school math courses until the 12th grade, when young women were slightly less likely than young men to enroll in a second-semester advanced math course. We used path analysis to determine whether this gender difference was mediated by constructs directly linked to expectations for success and subjective task value while controlling for the students' scores on the Differential Aptitude Test (DAT; Updegraff, Eccles, Barber, & O'Brien, 1996). As predicted, the gender difference in course-taking was completely mediated by these beliefs. Yet even more important, from my perspective, the gender difference was totally mediated by perceived importance or utility. It is interesting to note that neither 10th-grade enjoyment or interest nor self-concept of ability predicted number of courses taken once the DAT scores were included in the analyses.

We were particularly struck by the strength of the importance or utility construct. Recall the example I gave about the two young women deciding whether to take the college math course or not. I stressed there the perceived importance of the course for the young women's future plans. These data support that emphasis. At this point in these students' lives, they must begin to choose between elective courses. These findings suggest that they weigh the utility of the course for their future educational and vocational goals heavily in making these choices, and that gender differences in these course decisions are primarily due to perceived utility rather than either aptitude differences or differences in a sense of personal efficacy to succeed at mathematics. We found exactly the same pattern of results for gender differences in the number of high school physical science courses.

We next looked at the mediators of gender differences in career aspirations given their apparent role in course decisions. Here we used four sets of beliefs and values more directly related to career choices: (a) values regarding work, future success, relationships, and leadership (*lifestyle values*); (b) specific job characteristics adolescents may desire in their future occupational settings (*valued job characteristics*); (c) estimates of future success in different categories of occupations (*expected efficacy in jobs*); and (d) self-ratings of job-relevant skills (*self-perception of skills*; see Eccles, Barber, & Jozefowicz, 1999, for details on these scales and their psychometric properties). *Occupational aspirations* were assessed using the following open-ended probe: "If you could have any job you wanted, what job would you like to have when you are 30?" Analyses revealed fairly stereotypic gender differences: The young men aspired to science or math-related occupations, male-typed skilled labor occupations, and protective service jobs more than the young women. Conversely, the young women aspired to human service jobs, health professions, and female-typed skilled labor more than their male peers. However, the largest number of both males and females aspired to business or law occupations (31% and 30%, respectively).

We next analyzed which values, job characteristics, skills, and efficacy expectations best discriminated between adolescents who aspired to each of nine occupational categories (see Eccles et al., 1999, for full details). These analyses controlled for mathematical ability. Here I summarize the results for only two occupational categories: PhD-level or MD-level health careers and MS-level or PhD-level physical science-, engineering-, or math-related careers. First, as predicted by our model, both females and males who aspired to the health careers were more likely to expect to do well in health-related occupations and placed higher value on people or society-oriented job characteristics than those who did not aspire to health careers. The young women who aspired to these careers also rated their ability to succeed in science-related careers quite high as well. The young women, on average, were more likely to aspire to the health-related careers primarily because they placed higher value on a people- or society-oriented job than their male peers.

An even more interesting set of results emerged for the science-, engineering-, or math-related careers. Both the young women and men who aspired to these types of careers were more likely than their peers to expect to do well in science-related fields and to value math and computer job tasks. The young men who aspired to science-related careers also had high ratings of their computer and machinery skills and lower expectancies for doing well in business or law occupations than their male peers who did not aspire to these careers. Of interest, the young women who aspired to these types of careers placed much lower value on people- or society-oriented job characteristics than their female colleagues who did not aspire to these careers. This last effect was not evident for the young men.

These results are interesting for several reasons. First, they support the Eccles et al. model of achievement-related choices: For both males and females, occupational aspirations are mediated primarily by both expectancy beliefs and subjective task values. In addition, both approach-related (i.e., "I expect to do well in science, therefore I will choose a science career") and avoidance-related (i.e., "I do not value people- or society-oriented job tasks, therefore I will aspire to something else") beliefs predict the occupational choices for both males and females.

Second, there are intriguing gender and occupational category differences in the discriminating characteristics. For instance, expecting to do well in science-related occupations discriminates females who choose science-related or health careers from those who do not aspire to such careers. This is not true of males for whom only science-related expectancies discriminate between those males who choose science careers and those who do not. With regard to the females who choose science-related or health careers, it is important to point out that the value of people or society job characteristics also discriminates between those females who aspire to health, science, or math careers and those who do not. However, it discriminates in opposite directions for these two career options. That is, females who aspire to health careers place high value on people- or society-oriented job characteristics; in contrast, females who aspire to physical science-related careers place unusually low value on the people- or society-oriented aspects of jobs. Considering the fact that both groups of women expect to do well in science-related careers, it follows that one of the critical components influencing these women's decisions to go into a science- versus a health-related field is not science-related efficacy but the value they place on having a job associated with people and humanistic concerns. Thus, if we want to increase the number of females who consider entering physical science and engineering careers, it will be important to help females see that these careers provide opportunities to fulfill their humanistic and people-oriented values and life goals; such interventions are likely to be as successful, if not more successful than interventions designed to raise females' perceptions of their math-related abilities.

We have done comparable analyses on these young people's actual college majors and jobs at age 25 (Eccles & Vida, 2003). Again we sought to determine which beliefs and values distinguished those women and men who completed a physical science or engineering degree and who went into physical science and engineering careers from those who did not, controlling for actual mathematical ability. We found exactly the same pattern: As predicted by our model, both men and women who go into these professions have high expectations to succeed in these fields and place high subjective task value on doing the types of tasks inherent in these professions. Yet even more important, both the men and women who go into these fields place unusually low value on having a job that directly benefits other people or society. Although this effect is true for both men and women, the young women in this sample are much more likely than the young men to want jobs that provide direct benefits to society.

CONCLUSION

Our analyses suggest that the main source of gender differences in entry into physical science and engineering occupations is not gender differences in either math aptitude or a sense of personal efficacy to succeed at these occupations, rather it is a gender difference in the value placed on different types of occupations. Furthermore, our results suggest that these differences begin influencing educational decisions quite early in life. Finally, my own opinion is that these differences reflect, at least to some extent, inaccurate stereotypes about physical science and engineering that lead some young women and men to reject these careers for the wrong reasons. Many jobs in these fields do provide opportunities for individuals to fulfill humanistic and helping values. If we want to increase the number of females who aspire to and then actually go into these fields, we need to provide them with better information about the nature of these occupations so that they can make better informed decisions regarding the full range of occupations they might consider as they try to pick a career that fits well with their personal values and identity as well as their short- and long-term goals.

REFERENCES

Eccles, J. S. (1992). School and family effects on the ontogeny of children's interests, self-perceptions, and activity choices. In R. A. Dienstbier & J. E. Jacobs (Eds.), *Nebraska Symposium on Motivation: Vol. 40. Developmental perspectives on motivation* (pp. 145–208). Lincoln: University of Nebraska Press.

Eccles, J. S. (1994). Understanding women's educational and occupational choices: Applying the Eccles et al. model of achievement-related choices. *Psychology of Women Quarterly, 18*, 585–609.

Eccles (Parsons), J., Adler, T. F., Futterman, R., Goff, S. B., Kaczala, C. M., Meece, J. L., & Midgley, C. (1983). Expectancies, values and academic behaviors. In J. T. Spence (Ed.), *Perspective on achievement and achievement motivation* (pp. 75–146). San Francisco: Freeman.

Eccles, J. S., Barber, B., & Jozefowicz, D. (1999). Linking gender to education, occupation, and recreational choices: Applying the Eccles et al. model of achievement-related choices. In W. B. Swann, J. H. Langlois, & L. A. Gilbert (Ed.), *Sexism and stereotypes in modern society: The gender science of Janet Taylor Spence* (pp. 153–192). Washington, DC: American Psychological Association.

Eccles, J. S., & Vida, M. (2003, March). *Predicting gender and individual differences in college major, career aspirations, and career choice*. Paper presented at the biennial meeting of the Society for Research on Child Development, Tampa, FL.

Eccles, J. S., Wigfield, A., & Schiefele, U. (1998). Motivation. In W. Damon (Series Ed.) & N. Eisenberg (Vol. Ed.), *Handbook of child psychology: Vol. 3. Social, emotional, and personality development* (5th ed., pp. 1017–1095). New York: Wiley.

Jacobs, J. E., & Eccles, J. S. (1992). The impact of mothers' gender-role stereotypic beliefs on mothers' and children's ability perceptions. *Journal of Personality and Social Psychology, 63*, 932–944.

Ruble, D., & Martin, C. (1998). Gender development. In W. Damon (Series Ed.) & N. Eisenberg (Vol. Ed.), *Handbook of child psychology: Vol. 3. Social, emotional, and personality development* (5th ed., pp. 933–1016). New York: Wiley.

Updegraff, K. A., Eccles, J. S., Barber, B. L., & O'Brien, K. M. (1996). Course enrollment as self-regulatory behavior: Who takes optional high school math courses? *Learning and Individual Differences, 8*, 239–259.

III

CONCLUSION

ARE WE MOVING CLOSER AND CLOSER APART? SHARED EVIDENCE LEADS TO CONFLICTING VIEWS

STEPHEN J. CECI AND WENDY M. WILLIAMS

When we began this project, we hoped the top scholars' research would provide an answer to the question: Why aren't more women in science? We reasoned that if the dearth of women in some science and math fields could be attributed to gender differences in cognition, it would be an approachable task to distill experts' arguments and weigh the empirical evidence supporting rival claims. It is clear, however, that there are complexities that make it difficult to tally the pros and cons on a point-by-point basis. The way many of the arguments are framed makes it difficult to pit one type of evidence against another. Often, all sides in the debate draw on the very same evidence but interpret it differently. Below we outline five specific instances of this phenomenon, appraise the evidence supporting various interpretations, and, where possible, reconcile key questions. Next, we discuss a set of three interrelated issues: Are sex differences in cognitive skill large enough to matter? Are they inevitable? And, are they malleable? Throughout our discussion of these final three questions, we consider noncognitive factors as an alternative to cognitive ones for the dearth of women in some fields of science.

#1: The Case of the Right Tail of the Ability Distribution

All sides in this debate recognize that females, although roughly equivalent to males in *average* math and science test performance and slightly superior in some (but not all) verbal domains, are underrepresented at the extreme right tail of the SAT—Mathematics (SAT–M) score distributions—the top 1% or higher. Early ratios of males to females in the top 1% came from the Study of Mathematically Precocious Youth or SMPY (e.g., Benbow & Stanley, 1983), which favored males by nearly 12:1. However, ratios based on more recent testings have declined to as low as 3:1 in favor of males (Spelke, 2005). All sides agree that females score much lower than males on mental rotation tasks and on certain other spatial perception tasks that may be important in some science and engineering fields. A typical mental rotation task requires one to view two figures in different orientations and mentally rotate one onto the same plane as the other, or to determine whether a two-dimensional design on a sheet of paper can be folded into a three-dimensional shape. The size of the gender gaps on these types of tasks is the widest in the cognitive literature, with typically moderate to large magnitude gaps separating males and females. (In her review of 46 meta-analyses, Janet Hyde [2005] provided most of these effect sizes or magnitudes of the gender gap; see also a meta-analysis of 286 effect sizes by Voyer, Voyer, & Bryden [1995] for evidence that the magnitude of sex differences in spatial cognition has decreased in recent years on many but not all tests.) As noted below, we have no direct evidence that links this skill to specific job characteristics in fields in which women are underrepresented, such as computer science, physics, and engineering.

There is some evidence that sex differences in spatial cognition, including mental rotation, are evident prior to the onset of schooling. Several studies—but not all—have found significant male superiority in mental rotation among kindergartners (for a review, see Levine, Huttenlocher, Taylor, & Langrock, 1999). Such findings, if determined to be reliable, would not rule out an environmental basis for male superiority on mental rotation tasks, but they would prompt researchers to look elsewhere for the causes of sex differences among high school and college students that are often attributed to differential curricula, teacher reactions, and societal stereotypes. If the seeds of such sex differences are planted prior to educational experiences, the sources of such differences must be sought outside of strictly education-based settings. Of course, this all presumes that the evidence for sex differences in spatial cognition among 5-year-olds is robust and replicable.

Although girls score as well as or better than boys in elementary school science and mathematics, there is some slippage by high school, when fewer girls enroll in Advanced Placement (AP) chemistry and physics. Girls begin to score lower on some standardized tests of mathematics and science "aptitude" around this time, most notably the SAT–M, which allegedly is not closely tied to what is taught in the science and mathematics curriculum in high schools (a point made by Doreen Kimura in her essay in this volume [chap. 2]). All sides realize that, notwithstanding this "aptitude" gap among the top mathematics and science scorers, females achieve higher grades than males in most science and math courses, and they go on to receive math degrees at levels comparable with males.[1]

Diane Halpern (chap. 9, this volume) points out that this discrepancy between "aptitude" scores and classroom grades has led to claims of bias from both sides of the political spectrum. One side has argued that teachers and schools are biased against boys because they receive lower grades than their "aptitude" would appear to warrant. We hear also that schools fail boys by penalizing their impulsiveness and tendency to challenge teachers' authority. As one piece of evidence this side might summon, in a study of 67,000 college calculus students, males who received grades of D and F had SAT–M scores that were equal to females who received grades of B (Wainer & Steinberg, 1992). If this situation were transplanted to the case, for example, of a minority group whose IQ scores were high but who were given poor grades by their teachers, the inevitable criticism would be that the schools were failing these students. And yet, the other side of this debate might just as plausibly argue that it is not schools and teachers who are the culprits, but the standardized tests, which are biased against girls because they underpredict girls' grades in college calculus. As Spelke and Grace (chap. 4, this volume) comment in their essay,

> In U.S. high schools, girls and boys now take equally many math classes, including the most advanced ones, and girls get better grades. In U.S. colleges . . . men and women get equal grades, equating for educational institutions and mathematics classes. . . . Because males outscore females on the SAT–M, these findings indicate that the SAT–M systematically underpredicts the college mathematics performance of women, in relation to men. (this volume, p. 60)

When mathematical "aptitude" is equated by matching students on SAT–M scores, females outperform males in college mathematics courses, just as in the above calculus example invoked to claim bias against males (for a review, see Royer & Garofoli, 2005).

We have been placing quotation marks around the word "aptitude" because its meaning and measurement is a source of contention among psy-

---

[1]Currently, although only a small portion of all undergraduates major in mathematics, nearly 47% of them are females, and as Jacquelyn Eccles (chap. 15, this volume) notes in her essay, mathematics has long been regarded as a gender-neutral major. For example, in one SMPY cohort, 10.3% of men and 9.7% of women received bachelor's degrees in mathematics, and 2.2% of men and 2.1% of women went on to receive master's degrees in mathematics (Benbow, Lubinski, Shea, & Eftekhari-Sanjani, 2000).

chometric researchers, some of whom disagree that these tests actually measure underlying aptitude. For example, some argue that aptitude tests are not as impervious to environmental factors as was once thought (Ceci, 1996). This was why the Educational Testing Service changed the name of the test from *Scholastic Aptitude Test* to *Scholastic Assessment Test* several years ago. Many regard these tests as achievement tests—responsive to the number and level of mathematics courses enrolled in and to environmental factors such as encouragement—rather than as an index of native mathematics aptitude that is genetically programmed to flourish almost regardless of one's academic environment.

The resolution of this issue will have enormous bearing for the debate on women in science and math. If we imagine two bell-shaped distributions, one for native aptitude and the other for manifest performance on tests such as the SAT–M, the question becomes: Are those at the right tail in manifest performance there mainly because of their extraordinary *native aptitude*? Or is it possible to score in this manifest range possessing something less than extraordinary aptitude if one is exposed to all of the environmental conditions known to foster expertise (thousands of hours of practice, encouragement, role models, etc.)?

This distinction between *underlying aptitude* and *manifest performance* is at the heart of competing claims, even though it is rarely made explicit. One side suggests that it is possible to be successful in scientific, technical, engineering, and math (STEM) fields with a combination of *superior* (but not necessarily *extraordinary*) underlying aptitude, coupled with the right environmental conditions. If true, then there need not be any sex differences in these careers (i.e., if women themselves aspired to join such careers in greater numbers). This is because at the superior point in the manifest distribution (say, the top 5% to 10%), the sex imbalance is quite small—there are as many females who score in this region as there are males. The other side, however, argues that successful STEM scientists come from the *extraordinary talent* region of the aptitude distribution, say the top one tenth of one percent (0.1%) or even higher. In the remarks that inspired the debate that led to this volume, Lawrence Summers clearly came down on the latter side, noting that one is "talking about people who are in the one in 5,000, one in 10,000 class."

As further evidence, proponents of the "extraordinary talent" view can point to comparisons of the top quarter of the top 1% versus the bottom quarter of the top 1%. These elite groups both comprise individuals who would be regarded as very talented, but the top quarter of the top 1% comprises those persons whose ability exceeds 1 in 400, closer to the "extraordinary talented." Lubinski and Benbow and their colleagues (chap. 6, this volume) have reported these differences and, as we noted in the Introduction, the top quarter of the top 1% exceeds the bottom quarter of the top 1% on various outcomes, such as the number of tenured positions in top universities, number of patents, pay, age at promotion, and so on. This suggests that high manifest performance on tests such as the SAT reflects extreme aptitude as well as extreme supporting conditions—or else it is not easy to see why those in the bottom quarter of the top 1% would not fare just as well. Those in the bottom quarter of the top 1% seem just as motivated, as indexed by having taken as many AP courses and expressing equivalent aspirations in science and mathematics, for example.

Lubinski, Benbow, Webb, and Bleske-Rechek (2006) have extended this work. They reported that in a large sample of adolescents who scored within the top 1 in 10,000 in either mathematical (SAT–M $\geq$ 700) or verbal (SAT–V $\geq$ 630) ability before age 13, these individuals 20 years later had achieved tenure-track positions at top universities at rates comparable with graduate students attending the top U.S. math, science, and engineering doctoral programs. One would expect most of the graduate students at top universities to earn PhDs and go on to tenure track positions. But these extraordinarily high-scoring adolescents were far more elite than the graduate students and ended up being full professors more often then the graduate students (21% of extremely talented adolescents vs. 6.5% of former graduate students). As evidence for this assertion that the adolescents were more elite, their SATs placed them in the "1 in 10,000" region, whereas the mean GREs of the graduate students (quantitative scores in the mid-700s) placed them in the "1 in 100" region. Thus, in summary, the right tail of the ability distribution offers interesting insights about the causes of the dearth of women in science, but this research also raises many unanswered questions and is based on small, unrepresentative samples that may not reflect sex differences in the underlying ability as much as the vicissitudes of gathering nominations of mathematically precocious adolescent boys and girls, a point these authors note.

#2: The Case of Differing Gender-Related Real-World Demands

A second example of how both sides draw on the same or similar evidence but interpret it differently is provided by the claim that women are burdened by extra-academic demands that only rarely affect men, such as child rearing, caring for an elder parent, and so on. To some, such as Diane Halpern (chap. 9, this volume), this is evidence of an institutional barrier that has hindered women scientists from rapid promotion and advancement, particularly those with caregiving responsibilities, because of the additional requirements to put in lengthy laboratory hours. She opines that this is a better explanation for the underrepresentation of women in science than lower test scores. As we pointed out in the Introduction to this volume, there are ample survey data supporting her view. The National Science Foundation publishes every other year its Survey of Doctoral Recipients (SDR), a major statistical compendium of the characteristics of doctoral workers in science, engineering, and health fields. Data for the most recent SDR were collected over the period October 15, 2003 to May 31, 2005, on a nationally representative sample of about 40,000 U.S. residents who received doctorate degrees in science, engineering, and health from U.S. universities. The response rate was 80%. Doctoral degree workers were asked: "During a typical week on this [your principal] job, how many hours did you usually work?" Men and women without children reported working the same number of hours per week, 49.31 for men and 49.35 for women. However, women with one, two, or three children reported working between 1.7 and 2.5 fewer hours per week. Clearly, when children are involved, female scientists are doing the most caregiving, not male scientists.

To others, however, the point that women often engage in extensive caregiving amounts to an admission that extant salary and rank discrepancies are due to male scientists putting in longer hours and working more, uninterrupted by family demands. When women and men have similar cognitive profiles and put in similar hours, so it is argued, there are no salary or rank gaps between them. The *Science and Engineering Indicators* published by the National Science Board (2006) reports the results of a statistical technique called *linear regression*, which allows researchers to compare salary differences between men and women STEM scientists after they are adjusted to account for various factors. For example, scientists employed at large corporations and universities typically earn more than peers employed at small companies and colleges; scientists who have more experience (more years since receipt of highest degree) usually earn considerably more than those with less experience; scientists in some fields (computer science and engineering) usually earn more than those in other fields (life sciences and social sciences); and as we have already seen, the presence of children has a negative effect on women's hours per week but not on men's. The National Science Board findings are interesting because prior to adjustment, there are large salary differentials, with female PhD scientists earning nearly 26% less than male scientists. However, after adjustment to equate for prior experience, type of employer, children, and field, the salary differentials dropped to 3.1%.[2] In his incendiary remarks, Larry Summers invoked this potential real-

---

[2] These adjustments are somewhat controversial because the factors that are controlled are not randomly dispersed among men and women: Perhaps women do not enter higher paying fields because of a valid perception they are not wanted. Having said this, the use of such adjustments permits one to compare the earnings of men in a high-paying field with women in the same field as opposed to contrasting the salaries of women in a lower paying field with men in a completely different field, or, worse, comparing women with little experience with men with more experience. In the words of the report, "Salary differences between men and women reflect to some extent the lower average ages of women with degrees in most (STEM) fields. Controlling for differences in age and years since degree reduces salary differentials . . . (from 25.8% to –16.7%). . . . Controlling for field of degree and for age and years since degree reduces the estimated salary differentials for women to –10.3% at the doctorate level [reflecting the] greater concentration of women in the lower-paying social and life sciences as opposed to engineering and computer sciences. . . . Marital status, children, parental education, and other personal characteristics are often associated with differences in compensation. Although these differences may indeed involve discrimination, they may also reflect many subtle individual differences that might affect work productivity . . . most of the remaining salary differentials for

world distinction as a possible explanation for the dearth of women scientists in math-intensive careers: "what do we know, or what can we learn, about the costs of career interruptions . . . in what areas of academic life and in what ways is it actually true" (Summers, 2005).

Recent evidence from the Lubinski et al. (2006) survey reveals sex differences in how long men and women profess to work at their jobs while in their mid-30s. Consider Lubinski and Benbow's (chap. 6, this volume) argument:

> Regardless of the domain of exceptionality (securing tenure at a top university, making partner at a prestigious law firm, or becoming CEO of a major organization), notable accomplishments are rarely achieved by those who work 40 hours per week or less (Eysenck, 1995; Gardner, 1995; Zuckerman, 1977). World-class performers work on average 60–80 hours per week. . . .
>
> One only needs to imagine the differences in research productivity likely to accrue over a 5- to 10-year interval between two faculty members working 45- versus 65-hour weeks (other things being equal) to understand its possible impact. (p. 90)

Once again we see the same evidence interpreted differently. Is the problem that we lack a society in which sex roles are more nearly equal, where males share in child care and elder care, and where institutions are more aligned with the needs of female faculty throughout their life course? Some argue that no profession should expect of its members 60-hour work weeks that preclude doing much else, and that are associated with diminished satisfaction among professors. Others argue that very high levels of attainment require a level of continuous effort that would be unachievable if the sheer number of hours of effort was reduced. Answering such a question would seem to require a sensitive metric of effort that could be applied to STEM scholars at all ranks and pay grades. The best we can do at present is to rely on self-reported levels of work—not ideal, as psychologists know. Given this caveat, however, there are data that indicate that both sides may be right; that working longer is associated both with greater productivity and lesser satisfaction. Jacobs and Winslow's (2004) analysis of faculty satisfaction among a national sample of over 10,000 respondents revealed that faculty reported lower satisfaction as their self-reported work week got longer: 24% of men who reported working less than 50 hours per week were dissatisfied, a figure that jumped to 38.2% for men who reported working more than 60 hours per week. For women, the story was the same, with 30.3% of those working less than 50 hours per week being dissatisfied, a figure that jumped to 44.1% among those working over 60 hours. Unsurprisingly, mothers re-

---

women disappear when the regression equations allow for the separate effects of marriage and children for each sex. Marriage . . . has a larger positive association for men. Children have a positive association with salary for men but a negative association with salary for women." (National Science Board, 2006, Salary Differentials section, para. 4–12) ported the strongest dissatisfaction with working longer hours, and the highest dissatisfaction was found among those ages 35 to 54. And yet, this same survey showed that working 60 hours per week was associated with higher productivity. So there may not be an inconsistency in the claims that both sides make, because long work weeks are associated with both diminished satisfaction and higher attainment. Differing gender-related, real-world demands thus create the situation in which women work fewer hours at their career jobs, although they end up being more satisfied.

#3: The Case of Differing Gender-Related Preferences and Abilities

A third example in which both sides cite similar evidence is the now well-known litany of facts and figures showing that females tend to pursue people-oriented or organic professions (e.g., medicine, law, animal science, biology), whereas males with similar math and science ability profiles tend to pursue object-oriented fields (e.g., mathematics, engineering, computer science, and physics). But is this a bad thing? Not necessarily, we think. Fields have their own life courses, with their gender mix changing over time. Our own field (psychology) has become increasingly attractive to women to such a degree that most areas of psychology are producing more female doctorates than male ones, some by a ratio of 3:1. Although it would be deeply unfortunate if talented individuals were impeded from entering a field in which they were interested and capable of doing well, it strikes us as benign if females are, on average, interested in different fields than males—say, medicine as opposed to engineering—as long as their preferred fields are well regarded and well paid.[3] The most recent data on the share of PhD women occupying faculty positions in STEM fields show that their numbers increased sevenfold between 1973 and 2003, from 10,700 to 78,500, fully 30% of faculty posts and 28% of tenure track posts (National Science Board, 2006). It is important to note that if we look at the most recent cohort of female scientists, they occupy 40% of the junior faculty positions, which is almost exactly their share of that cohort's PhDs in STEM fields. Admittedly, this 40% share of new junior faculty is disproportionately in the social and life sciences.

Females enter biology, medicine, social sciences, and law in increasingly greater numbers, whereas their male counterparts with similar cognitive profiles enter computer science, physics, and engineering in greater numbers, sometimes by as much as a 3:1 ratio. Is it any less valuable for someone

---

[3]It is essential to take into account the very large pay differentials that currently exist at major research universities in recognition of the differential levels of competition from the private sector (e.g., economists earn more, on average, than other social scientists; computer scientists and engineers earn much more than humanists, on average, also because of market forces). Thus, if women are more interested in lower paying fields such as psychology, their salaries will be lower, just as male psychologists' salaries are lower.

in biology or medicine to be working on a cure for AIDS than for someone in engineering or mathematics to be working on a new search algorithm for Google? We think not. Essayists from very different perspectives (e.g., Eccles, Halpern, Kimura, Hyde, Hines, Lubinski, and Benbow) agree that we should explicitly add students' interests into the predictive mix, noting that males and females often have different interests that propel them into different careers.

In her essay, Melissa Hines (chap. 7, this volume) reminds readers that, historically, there have been pronounced swings away from male dominance in fields such as teaching, secretarial work, and medicine, and such shifts are easily explained in terms of changes in prestige, power, and income rather than by changes in hormones, cognitive abilities, or genes. Eccles (chap. 15, this volume) provides a very nice model validated by her longitudinal study, showing that young women were more likely than men to aspire to health-related careers primarily because young women place higher value on people- or society-oriented jobs than do their male peers (and this remained true even when their mathematical ability was taken into account).

David Geary (chap. 13, this volume) suggests that such differential interests are to some degree the modern vestige of patterns of intrasexual competition or mate choice. Ruben and Raquel Gur (chap. 14, this volume) suggest that differential interests are due to some other force, such as an evolutionary male advantage in actions requiring rapid transition from perception (posterior) to action (anterior), incorporating limbic (ventral) input. Alternatively, differential interests among males and females may result from more proximal societal forces. This is clearly an interesting theoretical issue because its resolution would need to take into account the neurobiological differences that the Gurs and others note, while simultaneously taking into account the secular trends in gender shifts in jobs that Hines notes. But, for now, our view is that implementing incentives to encourage greater female participation in nonpreferred fields could risk females' finding such professions to be less exciting and less rewarding, leading ultimately to lowered rates of success. The one major caveat to this assertion is the observation by Eccles (chap. 15, this volume) that high school students (and we would add college students as well) often possess erroneous stereotypes of the career options available in fields such as engineering, physics, and computer science. It is one thing to be disinterested in a career for valid reasons; it is quite a different matter to be disinterested in it for inaccurate reasons. This is one type of intervention that ought to be easy to offer: Expose students to information about a range of career options in STEM fields so they fully understand the career possibilities when they begin their implicit cost–benefit analysis of which careers are worth what level of effort and delayed gratification. Thus, in conclusion, differing gender-related preferences are clearly an important component of this debate, although the source of these differences is not completely clear at the time of this writing.

#4: The Case of the Interaction of Biology and Environment

Despite some disagreements among researchers about the role of biology in gender differences, there is an acknowledgment by all sides in the debate that biological potential is mediated by the environment. What this means is that a biological or genetic factor may create the possibility for some behavior, but whether the possibility will materialize depends critically on the environment.

Take the case of prenatal hormones, a topic discussed by those on all sides of this argument (see essays by Hines [chap. 7], Berenbaum & Resnick [chap. 11], and Kimura [chap. 2], this volume). Putting aside the specifics of their arguments, all would agree that excessive levels of androgen are associated with masculine behaviors on the part of girls, most vividly illustrated in those with congenital adrenal hyperplasia (CAH). Yet, CAH girls may end up with superior spatial ability and male behavior patterns that may or may not be due to the direct action of the hormone. As Berenbaum and Resnick note, the role of the environment cannot be ruled out, an argument accepted by all. Perhaps CAH girls are treated differently as toddlers, evoke different responses from others, are selected into male-typical activities because of their greater masculinization, and end up being better at such activities— not directly because of androgen but indirectly because of the reactions from others.

Dickens and Flynn (2001) termed this impact on development a *multiplier effect*: Initial small behavioral differences may evoke responses that multiply the initial state or advantage, resulting in an upward-cascading sequence of interactions that eventuates in very large differences during adolescence. Dickens and Flynn showed mathematically how multiplier effects can account for large cognitive differences, despite small initial differences. They explained the seeming paradox of high heritability with large environmental influences. As an example of someone with a small initial advantage who ends up with a large advantage because of the environment, they describe the case of athletic prowess. An individual gets noticed for slightly superior initial ability handling a basketball. People comment on it, causing the individual to focus on it; teams choose this individual more often, he or she gets better coaching, more competition, more praise, and so on. The result is an advantage during adolescence that far exceeds any initial biological advantage.

The point is to make clear that environments are "biologically loaded," making it difficult to disentangle the relative contributions of hormones and environment to outcomes such as spatial ability and career aspirations. Thus, biological factors may prod women toward certain careers not because of the direct effect of biology on cognitive ability, but rather through the indirect effect on reactions that prod one toward certain aspirations. Socialization theorists would agree with this point, and so would biologically oriented theorists.

#5: The Case of Greater Male Variability

A final example of the same evidence being interpreted differently concerns the claim that males exhibit greater variability, leading to an overrepresentation of extreme scores (both very low and very high). All sides in the gender wars agree that there is greater variability in male distributions of many abilities, and that this sometimes results in more males at the very lowest and very highest ends of the distribution. To some, this helps explain why there are so many more male engineers, mathematicians, and physical scientists. This is because even if one assumes that the abilities associated with sex differences at the right tail are only modestly important to success in those fields of science in which women are underrepresented, and even if one assumes that the sex ratio is small at the right tail (say, 2 to 1 in favor of males), these factors could still explain larger pools of males vying for professional positions.

In the SMPY program, after screening large numbers of boys and girls (note that the cutoff for admission was more elite than the 1% level at which the sex disparity is greatest), those admitted were given accelerated exposure to mathematics. At the end of high school, these students took the SAT–M again as part of the process of applying to college, and again there was a preponderance of boys at the upper right-hand tail of test scores (Benbow & Stanley, 1983). The investigators concluded that there were more boys than girls in the pool from which future scientists and mathematicians are drawn. Because the initial difference was obtained before students began to select their courses and because the students showed few sex differences in their reported attitudes toward mathematics, the investigators suggested that the sources of the sex difference were, in part, genetic (Benbow, 1988; Benbow & Stanley, 1983; see also Pinker, 2002). In her essay, Eccles (chap. 15, this volume) also notes the appearance of sex differences in interests very early, though her interpretation of this is not biological.

To others, however, the very same data do not compel an account in terms of biologically based sex differences. After all, if the genetic contribution were strong, then males should predominate at the upper tail of performance in all countries and at all times, and the male–female ratio should be of comparable size across different samples. None of this seems to be the case (Feingold, 1992; Shayer, Ginsberg, & Coe, in press; see also Spelke, 2005, p. 956).

ARE SEX DIFFERENCES INELUCTABLE?

Sex differences appear to be neither as unambiguous as earlier researchers suggested nor as insubstantial as some current critics claim. Sex differences in career choices are definitely not inevitable, as the past 30 years have documented a sea change in the gender makeup of various fields. There are many interesting ecological niches in which sex differences, at least around the midpoint of the distribution, do not exist, or favor females (see Shayer et al.'s [in press] finding with the Volume and Heaviness test, in which a very large male advantage in spatial reasoning essentially disappeared over the past 30 years among British adolescents as a result of a large drop in performance by males). In her essay, Newcombe (chap. 5, this volume) describes successful interventions that have elevated girls' spatial skills, even if the gender gap does not completely close. Among low-socioeconomic-status (SES) third graders, girls and boys do not differ notably in spatial skills, and middle-class girls are at least as good as lower-class boys on these tests (Levine, Vasilyeva, Lourenco, Newcombe, & Huttenlocher, 2005). Among some Eskimo groups in which females and males both hunt, there is no significant spatial skill gap. Icelandic high school girls are actually superior to boys on spatially loaded subtests (Levine et al., 2005), and Japanese girls are far superior to American boys on math scores, with girls in Singapore excelling over everyone, often by a full standard deviation!

In fact, the cross-cultural findings are rife with examples of females in some nations (Taiwanese, for example) vastly exceeding American and Canadian males (see Valian and Hyde essays in this volume [chaps. 1 and 10, respectively]). This evidence all points to the non-inevitability of spatial gaps, at least when we talk about average numbers of correct answers, performance in high school mathematics classes, and the percentage of females pursuing math and science majors in college (see Spelke & Grace, chap. 4, this volume). When we talk about performance at the extreme right tail of the distribution, the picture is not always as clear-cut. Here males are overrepresented, but the degree of their overrepresentation depends on the measure of ability used and the cohort studied, as Janet Hyde points out in her meta-analysis.

As far as the remediability of cognitive skills is concerned, there are fascinating data showing that girls and boys often attempt to solve speeded test problems differently, but when they are instructed to use the same strategies they can in fact do so. In other words, females, while preferring certain strategies that are nonoptimal on speeded tests such as the SAT–M, can use optimal strategies when instructed to do so. This suggests that females have the cognitive *ability* to use the best strategies and score optimally but choose, for whatever reason, to use less optimal strategies.

In addition to direct training interventions, one often overlooked but potentially important intervention is to provide all students with accurate information about various STEM careers. In her essay, Eccles (chap. 15, this volume) notes that there is a "disconnect" between the options a field offers and students' beliefs about these options. Better information might go a long way toward encouraging cognitively competent students of both genders to pursue careers they are capable of doing well in, and receive enjoyment and satisfaction from roles they are unaware of in those careers.

One last word on the question of remediation through training: Ceci and Papierno (2005) reviewed the training literature with an eye toward closing the gender gap. They reported that sometimes training interventions actually widened preexisting gaps—if these interventions are made available to all students, not just to those in greatest need. This is because it is sometimes the case that the biggest gains in training studies are made by those who were the highest scoring before the intervention. The idea of targeting training interventions to one sex and excluding the other from its potential benefits raises many interesting political, economic, and moral questions that are beyond the scope of this volume (see Ceci & Papierno, 2005; Papierno & Ceci, 2005). But in the context of the present debate, it bears noting that some interventions designed to elevate the numbers of women in STEM careers (e.g., educating high schoolers about the actual possibilities that each field offers to counter their erroneous stereotypes, or training adolescents on spatial reasoning skills) could end up increasing the pool of females, but increasing the pool of males as much or more so. These are empirical questions worthy of study. Baenninger and Newcombe's (1989) meta-analysis showing that spatial ability is malleable (see below) elevated boys as much as girls; hence it did not affect the preexisting spatial gap. And the spatial training designed by Vasta and Gaze (1996) totally eliminated one form of spatial gap and reduced the magnitude of another, but in neither case did this training transfer to related types of tasks that were not directly trained.

ARE SEX DIFFERENCES LARGE ENOUGH TO MATTER?

The answer to the question posed in the heading is "it depends." As we noted in the Introduction, it depends on what we are trying to predict, whether it is under existing conditions, and how much we are willing to alter the situation to elevate female spatial and mathematical ability. Regarding the first, if we are trying to predict success at most jobs, college admission decisions, or just about anything that does not require an extraordinary level of performance on spatial or speeded mechanical ability tests, the magnitude of typically observed sex differences should not lead to important differences in predictions. Most individuals of both sexes have the necessary spatial ability to perform well. Furthermore, if girls and women are instructed to use strategies that are optimal for speeded tests, they appear to benefit significantly.

What we do not know is the extent to which certain fields of science require spatial cognition of the absolutely highest degree. Do some fields rely on being in the top 0.5% in the ability to visualize n-dimensional space, or rotate drawings into three-dimensional (3-D) pairs or their mirror images? Or is there little gain accrued from performing above the top 25% on such skills (a group in which there is little or no gender asymmetry)? We do not know. Hence, it does seem that the ability to represent and transform spatial information is an important component of such activities as navigating from maps and reading graphs, architectural drawings, and X-rays. Such tasks often require the ability to mentally transform images and reconstruct 3-D forms from two-dimensional images, but there is no evidence directly linking measures of spatial reasoning with success at these activities.

Even if males outnumber females at the extreme high end of the SAT–M distribution (Benbow, 1988), females have demonstrated very significant success in graduate programs in science and engineering, more than quadrupling their numbers in the past 40 years in the United States. We know also that spatial cognition can be enhanced through direct training (e.g., Vasta & Gaze, 1996). That schools do not train such skills could be changed if it turns out that these skills are critically important for STEM professions and that such training generalizes to closely related forms of spatial cognition. Hyde's recommendation in her essay (chap. 10, this volume) that all college-bound students be required to take 4 years of high school math and 4 years of high school science strikes us as sensible. But it will only work if colleges and universities make this type and level of preparation a condition of admission.

Nora Newcombe (chap. 5, this volume) and Melissa Hines (chap. 7, this volume) both remind readers that there are many factors influencing success at the highest levels of science, in addition to spatial and mathematical skill:

> Beyond some (high) threshold, I doubt that extra increments of the same cognitive ingredients explain much variance. Thinking creatively, explaining one's data, or inspiring a research team may be pretty important as well! . . . If we want to maximize the human capital available for occupations that draw on spatial skill, such as mathematics, engineering, architecture, physical science, and computer science, we would do better to concentrate on understanding how to educate for spatial skill rather than focus solely on the explanation of sex differences. (Newcombe, this volume, p. 75)

Of course, this was not the context for Summers's remarks; he was interpreted by some as suggesting that the dearth of female scientists, mathematicians, and engineers may be due to cognitive causes that are biologically rooted. In Hedges and Nowell's (1995) analyses of probability samples published between 1960 and 1994, the magnitude typically observed was a 0.5 standard deviation sex difference on tasks that require rotation in mental space of 3-D objects or visualization of spatial relationships—large enough to matter if being in the top 1% or higher on those skills is critically important for success. A moderately large gap of this size between males and females at the center of a distribution of scores on spatial ability will translate into many more males at the tails of the distribution.

Consider an example: Assuming roughly equivalent standard deviations among male and female samples (actually males have somewhat larger variance for mathematical and spatial tests, usually around 10% to 20% larger than female variance), homogeneity of variance, bivariate normality, and roughly equal sample sizes, if we randomly select male and female scores from a combined sample of male and female children, this will result in the male child slightly exceeding the female child. However, the biggest impact of such a gap will be observed when the selection criterion is moved up—when we attempt to select only elites (or, conversely, only the poorest performers). For example, if only the top 5% of students in a class that is 50%–50% male–female is eligible for gifted and talented programs, such programs will have many times more males than females if there is a 0.5 standard deviation gap separating the sexes at the midpoint of the distribution.

If we look at the top 1% of the mathematics distribution, males will outnumber females by approximately 7 to 1. This is an outcome of a normal distribution of scores with greater variance for male scores; in such cases even small differences at the mean of the distribution can become much larger differences at the tails of it. In the context of Summers's comments, it would mean that far fewer women score in the top 1% than men, the part of the spatial ability distribution in which most scientists presumably reside. This point was made by others (e.g., Hedges & Nowell, 1995): "Differences in the representation of the sexes at the tails of the ability distribution are likely to figure increasingly in policy about salary equity" (p. 41).[4]

However, this prediction regarding membership in the far right tail of the distribution and likelihood of entering and succeeding in STEM professions presupposes the assumption we have been making: namely, that there is a direct relationship between cognitive ability and career choice and success. A number of the essayists call this assumption into question (e.g., see Halpern's essay, chap. 9, this volume). This prediction also assumes that the measures we use to assess sex differences are themselves valid in the sense that they truly tap the skill in question. We turn to these two assumptions next.

The assumption of a direct causal link between cognitive scores and success in one's chosen scientific career ignores the reality of the tenure system. Once a woman is hired, the tenure system—with its requirement to show excellence at a young age and its insistence on full-time employment—make it particularly difficult for women (who generally do the major child-rearing and elder care and, although this is not a topic that many find politically appealing, often defer their careers to those of their male partners). None of these factors make universities easy workplaces for women, even when progressive family leave policies are in place.

---

[4]For reading comprehension, perceptual speed, and associative memory, females outnumbered males in the top 5% and 10% of performance, and males were 1.5 to 2.2 times as likely as females to score in the bottom 5% and 10% of the score distributions. For both spatial reasoning and mathematics, males were between 1.5 and 2.3 times more likely to be at the high end of the score distribution (including 7 times more likely to be at the top 1%). Where males were hugely overrepresented at the high end were in areas of mechanical/electronic reasoning (by a factor of 9 to 10). It is interesting to note that there are overrepresentations of males in social studies by 1.7 to 3.5, which is rather odd, given the verbal nature.

However, before we assume that enacting even more progressive policies would change the representation of successful women in STEM careers, a great deal more data are needed. Can scientists be as successful if they reduce their scientific effort for a period of time while they rear children and care for elderly parents, especially if this reduction occurs early in their careers? If universities and colleges offered variations on the rigid lockstep tenure clock, what implicit disincentives might cause women not to take advantage of them, or for men to "game" the system to their advantage? And if women take advantage of such variations, what would the effect be vis-à-vis their colleagues who worked uninterruptedly full time?

The second assumption is very difficult to test because it goes to the heart of construct validity, or how and whether a test measures what it claims to measure. The political scientist James Flynn, whose work on intellectual trends over the 20th century is known as the *Flynn effect*, has shown that Jewish Americans score somewhat lower on spatially loaded test items—and yet they are, if anything, overrepresented among scientists and mathematicians (see Flynn, 1991, pp. 119–123). In light of such findings, it seems hard to argue that female scores on spatial reasoning account for their low numbers in scientific and technical careers (Flynn, 1998). At the very least, it is hard to argue that this is the only or the major factor.

This raises the issue of whether the instruments we use are accurately assessing the abilities we think are important to success in STEM fields. The SAT–M is the most widely used index in the debate over sex differences, but it was never designed to answer this question. We know little about whether the questions asked on it are the ones that truly tap mathematical abilities needed for success in STEM careers. This is not to say that we know nothing about its predictiveness of college students' grades in math and science courses; the SAT–M does modestly predict math grades in college and, when it is added to other information such as values and preference measures, it improves the prediction even more. We also know its link to other abilities such as mental rotation.

For example, Casey, Nuttall, Pezaris, and Benbow (1995) found that the sex difference on the SAT–M was eliminated in several samples when the effects of mental-rotation ability were statistically removed. This suggests that rotational skill may mediate certain high-level mathematical abilities, or at the least, that these two abilities tend to covary. Importantly, the relation between spatial cognition and mathematics was found to be much stronger for females than males, suggesting that females may be particularly hindered by their spatial skill. On the other hand, many measures are correlated with the SAT–M and also with college grades in math and science, but they have nothing to do with mathematical ability (e.g., parental education and income, race, and birthweight). (For example, see Ceci, 1996, for data showing a 0.4 correlation between parental SES measures and children's SAT scores.) But we would not leap to the conclusion that these variables predict college grades in math because they are measures of underlying mathematical talent. That argument becomes circular because the sole touchstone is the SAT–M score—and it cannot be used to validate itself. Parental social class variables may be proxies for a stream of unmeasured conditions that are related to achievement.

What this means is that the SAT–M cannot tell us whether scoring in the top 1%—as opposed to, say, scoring in the top 15%—taps the skills needed to be a successful scientist in a mathematically intensive field. Are there specific items that must be answered correctly to be successful in STEM careers? Are they the ones that tap spatial abilities involved in mental rotation? What variance in scientific success do such skills account for when we add to the mix measures of creativity, diligence, risk taking, other forms of mathematical skill, communicative ability, and personal preferences? We do not know. Finally, we do not know what portion of women are capable of succeeding in scientific careers but fail to persist in their training because of personal reasons that have nothing to do with cognition, a point made by Diane Halpern (chap. 9) and others as well.

The most thoughtful analyses that bear on this question have been conducted by Wai, Lubinski, and Benbow (2005) and by Achter, Lubinski, Benbow, and Eftekhari-Sanjani (1999). Both of these longitudinal studies contrasted two groups of extremely talented adolescents—those who scored in the top quartile of the top 1% versus those who scored in the bottom quartile of the top 1%. Their results show that those who as adolescents scored in the top quarter of 1% when studied 10 to 20 years later received significantly greater numbers of PhDs in science, were more likely to be tenured at top universities, and were credited with more inventions and patents. This would seem to indicate that those who are successful in STEM fields are disproportionately likely to come from the very top of the distribution. But because of the sample size and greater variability among male scores, the findings cannot readily be applied to sex differences.

Another way to look at this issue is to ask what each item on the SAT–M measures and how each item relates to later success. Spelke (2005) described a case in which changes made to items on the SAT–M could have resulted in bias against women—or perhaps not, depending on information that we do not know.

Using Spelke's (2005) example, girls consistently outperform boys on so-called data sufficiency problems. These are items for which the student must determine if the data provided in a problem are sufficient to answer the problem. These items have been removed from the SAT–M because they are known to benefit from coaching. However, as Spelke noted, removing a class of items on which girls score better will of necessity lower their scores relative to boys'. Will the removal of such items increase or decrease the fairness of the SAT–M as a measure of women's mathematical ability?

If boys are more talented than girls, then this change may have increased the fairness of the test. If boys and girls are equally talented, then this change increased the test's bias against girls. Evaluation of the SAT–M therefore requires an independently motivated account of the nature of mathematical talent, its component processes, and its distribution across boys and girls. (Spelke, 2005, p. 954)

We simply do not know the extent to which tests such as the SAT–M assess the skills critical for STEM field success, notwithstanding their modest correlations with college grades and other measures. Such tests may "proxy" for unmeasured variables that, even if they contribute to the prediction of mathematics grades, are not of necessity key aspects of the mathematical skills needed to be a successful scientist. It would be informative if there was some agreed-on measure of underlying mathematical aptitude that was known to tap the same talents (or their precursors) deemed critical for success in mathematically intensive careers, and which was not influenced by the social variables that numerous contributors alighted upon (e.g., stereotype threat). The SAT–M may in fact be such a measure, but until we have better construct validation for it as a measure of underlying aptitude, there is a circularity inherent in using it to both predict sex differences, on the one hand, and validate sex differences in underlying mathematical aptitude, on the other.

ARE SEX DIFFERENCES MALLEABLE?

There is a long-standing literature showing that virtually all higher level cognitive abilities (ones requiring thinking and reasoning that take place on a conscious level of awareness) are affected by experience (Ceci, 1996). Spatial and mathematical abilities are no exception, and in her essay in this volume Nora Newcombe (chap. 5, this volume) describes some of the recent interventions that have been shown to elevate female spatial skills, though not to a degree that completely closes the gender gap, she notes. Her own pioneering meta-analysis showed that experience can alter spatial ability (Baenninger & Newcombe, 1989). A more recent study showed that the difference could be eliminated by carefully conceptualized training (Vasta & Gaze, 1996). In that study, spatial training of college students on the water level task, using a self-discovery training procedure with progressively more difficult tasks, was effective in totally eliminating preexisting gender differences on a spatial drawing task and significantly improving females' knowledge of the spatial (invariance) principle, although not to the level of males.

Factors such as educational experiences (highest grade attained, number of science and math courses taken, availability of AP courses), social class (Levine et al., 2005), recreational play, and outdoor exploratory behavior (Entwisle, Alexander, & Olson, 1994) have all been implicated in the development of spatial ability, at least to some extent. In her essay, Carol Dweck (chap. 3, this volume) provides fascinating evidence that the mathematics gap that begins to emerge in junior high school can be closed by carefully scripting the messages we send girls. Her demonstrations have the advantage of taking into account any differences in mathematical scores that existed before the start of her interventions, thus excluding alternative explanations.

Of course, none of this rules out the possibility of initial biological differences between males' and females' spatial abilities that propel relatively more males toward such activities. As Doreen Kimura (chap. 2, this volume) notes, male rats are superior to female rats in learning spatial mazes, and these sex differences can be reversed by hormonal manipulation in early postnatal life. This suggests that the social environment may not be the cause of gender differences, even if it can be the remedy. Nor do such findings say anything about the neurobiological consequences of engaging in such experiences (e.g., Halpern's [chap. 9, this volume] point that certain activities may lead to changes in the brain). Perhaps males have a greater propensity to engage in spatially loaded games (e.g., Legos, Rubik's Cube, puzzles, building modular racer cars online) and explorations of their neighborhoods, and to take more physics courses—even when such toys and experiences are available to all—with the consequence that such activities promote brain growth. But these types of findings do make it clear that achievement on spatial tests can be influenced, often highly.

As Nora Newcombe points out in her essay (chap. 5, this volume), the magnitude of the spatial gains due to training is often larger than the magnitude of sex differences. And for those with a genetic bent, there is one finding that presents a special problem: Throughout the second half of the 20th century, spatial ability has increased faster than the gene pool can be expected to have changed. Consider that James Flynn has repeatedly shown that so-called "fluid" intellectual abilities, of which those that involve spatial skills are the most prominent, have grown faster than all other abilities (Flynn, 1987). The Raven's Progressive Matrices, a seemingly spatial reasoning test,[5] shows the largest gains over the 20th century, about two to three times larger than

---

[5]Among the myriad complexities of gender differences in cognitive abilities is the issue of what to call the abilities. For example, the Raven's is a well-known test that involves presenting increasingly complex visual arrays with sections missing. The challenge is to select the missing swatch from among the options offered, a type of visual multiple-choice test. On its face, the Raven's would seem to be a paradigmatic test of sequential, static spatial abilities. However, among psychometric researchers the Raven's is regarded as primarily a test of general intelligence (g) and only secondarily a test of spatial visualization (see Carroll, 1997). The same is true of other measures that have been invoked in the debate over sex differences in cognitive abilities. The reason this may be more than an academic matter is that although 25 years ago women performed notably worse than men on the Raven's, they now appear to perform as well or better. If it is primarily a test of g and only slightly a test of spatial visualization, then this can be explained in terms of the Flynn effect, whereas if it was really measuring spatial ability, then it would represent startling evidence that women are as good or better than men on spatial skills tapped by this type of test.

gains for nonspatial abilities such as vocabulary and verbal reasoning. And, if we examine the subtests of the major intelligence tests such as the Wechsler series, we find that two of the top five gainers between 1948 and 2002 are heavily spatial (Block Design and Object Assembly). Performance on each of these has escalated dramatically over the 20th century, often by over 1.5 standard deviation. If spatial ability is under genetic control, it is not clear how such enormous gains could have occurred over such a brief epoch. Although it is true that none of these tests tap mental rotation per se, they are undeniably spatial, and the Block Design test has some of the same ingredients as mental rotation tasks.

Having said that cognitive performance can be altered by training, we note a fascinating complication: Ruben and Raquel Gur (chap. 14, this volume) provide highly suggestive evidence that the brains of males and females are organized differently. Specifically, in remarking on observed sex differences in the activation of different brain regions while solving a 3-dimensional spatial maze, the Gurs note that females appear to draw on more parietal and prefrontal regions, whereas males rely more on the hippocampus structure. The reason this is interesting is because it would appear that females are engaging in a conscious, deliberate form of processing while performing spatial cognition tasks (the prefrontal region), whereas males seem to be relying on an automatic processing of geometric-spatial cues (also see Grön, Wunderlich, Spitzer, Tomczak, & Riepe, 2000). If it were the other way around, it would be easier to imagine how training studies could affect female performance; females could be trained to engage in the same deliberate processes that males use to solve these spatial tasks if males solved these problems by drawing on conscious prefrontal resources. However, the fact that males appear to be engaging in automatic forms of processing suggests that it will be more difficult to train females to emulate males. Of course, there may be multiple routes to improving performance, and perhaps ones that involve different neural routes compared with those used by males may prove fruitful. But this cautionary note deserves mention because it could explain why even the most carefully crafted interventions that have been cited to document the malleability of spatial ability have sometimes failed to produce transfer from the specific performance trained to a highly related one that was not directly trained (Vasta & Gaze, 1996).

We conclude this volume by considering a knotty dilemma: Do we vocally embrace the study of gender gaps, or do we pursue such study (if at all) quietly, framing questions in a manner that keeps the lay public from grasping the societal implications, especially if it should turn out that some of the gender asymmetry is the result of biologically based cognitive differences—though we are not proposing that this is the best interpretation of the data at this point? As scientists who spend their professional lives studying gender differences in math and science achievement, the essayists are aware of the dangers associated with sending girls the wrong message. Dweck's work speaks to this problem directly (chap. 3, this volume), but so do many others less directly. If parents, their daughters, and those who teach them hear about research showing gender gaps at the right tail of the bell curve, might this convey to them that math, science, and engineering are not "their" bailiwick? Do we want to communicate that engineering and computer science are for boys, and that no matter how the sexes do on average, boys still rule at the right tail, which is where scientists come from? All of us would agree that great sensitivity is needed in the way we design our studies and discuss our findings so that we do not end up hurting females' prospects unintentionally, although the already high numbers of females doing well in mathematics and some fields of science and medicine suggest this is not a major part of the problem. As Berenbaum and Resnick (chap. 11, this volume) note, parental attitudes and media portrayals of sex differences in ability may affect individual characteristics related to career choice (e.g., by moderating or even mediating the effects of sex hormones).

Having said this, we can all point to historical examples in which muting of scholarship because of social or political reasons has set back science, or sent a chill through some camps doing controversial research.[6] As scientists, we should all recognize the importance of free and open inquiry, absent subtle (and perhaps not-so-subtle) messages that one side (e.g., those who are persuaded that gender differences are rooted in brain differences) is offensive. Sadly, this is the state of affairs for many politically charged topics (e.g., group differences in intelligence, the effect of elective abortion on women's mental health outcomes, etc.), and it probably stymies progress in thinking as well as in public policies, as Doreen Kimura argues in her essay (chap. 2, this volume). All legitimate views need to be aired openly for science to flourish and for policies to be well informed by scientific findings.

But we also need to be careful that we put forward our views in a way that does the least harm, because there is danger that the research will be translated into self-handicapping messages by females, as Carol Dweck's (chap. 3, this volume) studies demonstrate and as Berenbaum and Resnick (chap. 11, this volume) suggest. If there should ever emerge a consensus in the neuroscience research community that the biological underpinnings for very high levels of mathematical and spatial reasoning are not as common among girls as boys, someday it may still be possible to ameliorate the disparity, much in the way neuroscience has done for Alzheimer's disease, schizophrenia, and other deficits once thought to be impervious to intervention.

It is a knotty problem, indeed, and we can do no better than conclude with the words of Richard Haier (chap. 8, this volume):

---

[6]The case in Britain of Frank Ellis at Leeds University is a recent example. Ellis was suspended by his university's vice chancellor for stating that racial differences in IQ are a full standard deviation and speculating about biological bases of these differences, even though his comments were drawn from Herrnstein and Murray's (1994) *The Bell Curve*, which was also condemned as hateful speech not covered by academic freedom.

The challenge is to follow where the data lead, always cognizant of Orwellian fears and prejudiced misuse of knowledge balanced by the prospects of alleviating suffering from disorders and enhancing the quality of life for everyone. Along the way, controversy can only escalate as we constantly test new knowledge against old and comfortable ideas. This is the way science works and the way our culture evolves. (p. 119)

REFERENCES

Achter, J. A., Lubinski, D., Benbow, C. P., & Eftekhari-Sanjani, H. (1999). Assessing vocational preferences among gifted adolescents adds incremental validity to abilities. *Journal of Educational Psychology, 91,* 777–786.

Baenninger, M. A., & Newcombe, N. (1989). The role of experience in spatial test performance: A meta-analysis. *Sex Roles, 20,* 327–344.

Benbow, C. P. (1988). Sex differences in mathematical reasoning ability in intellectually talented preadolescents: Their nature, effects, and possible causes. *Behavioral and Brain Sciences, 11,* 169–183.

Benbow, C. P., Lubinski, D., Shea, D. L., & Eftekhari-Sanjani, H. (2000). Sex differences in mathematical reasoning ability: Their status 20 years later. *Psychological Science, 11,* 474–480.

Benbow, C. P., & Stanley, J. C. (1983, December 2). Sex differences in mathematical reasoning ability: More facts. *Science, 222,* 1029–1031.

Carroll, J. B. (1997). Psychometrics, intelligence and public perception. *Intelligence, 24,* 25–52.

Casey, M. B., Nuttall, R., Pezaris, E., & Benbow, C. P. (1995). The influence of spatial ability on gender differences in math college entrance test scores across diverse samples. *Developmental Psychology, 31,* 697–705.

Ceci, S. J. (1996). *On intelligence: A bioecological treatise on intellectual development.* Cambridge, MA: Harvard University Press.

Ceci, S. J., & Papierno, P. B. (2005). The rhetoric and reality of gap-closing: When the "have-nots" gain, but the "haves" gain even more. *American Psychologist, 60,* 149–160.

Dickens, W. T., & Flynn, J. R. (2001). Heritability estimates versus large environmental effects: The IQ paradox resolved. *Psychological Review, 108,* 346–369.

Entwisle, D. R., Alexander, K., & Olson, L. S. (1994). The gender gap in math: Its possible origins in neighborhood effects. *American Sociological Review, 59,* 822–838.

Feingold, A. (1992). Sex differences in variability in intellectual abilities: A new look at an old controversy. *Review of Educational Research, 62,* 61–84.

Flynn, J. R. (1987). Massive IQ gains in 14 nations: What IQ tests really measure. *Psychological Bulletin, 101,* 171–191.

Flynn, J. R. (1991). *Asian Americans: Achievement beyond IQ.* Hillsdale, NJ: Erlbaum.

Flynn, J. R. (1998). WAIS–III and WISC–III: IQ gains in the United States from 1972 to 1995; how to compensate for obsolete norms. *Perceptual and Motor Skills, 86,* 1231–1239.

Grön, G., Wunderlich, A. P., Spitzer, M., Tomczak, R., & Riepe, M. W. (2000). Brain activation during human navigation: Gender-different neural networks as substrate of performance. *Nature Neuroscience, 3,* 404–408.

Hedges, L. V., & Nowell, A. (1995, July 7). Sex differences in mental test scores, variability, and numbers of high-scoring individuals. *Science, 269,* 41–45.

Herrnstein, R. J., & Murray, C. (1994). *The bell curve.* New York: Free Press.

Hyde, J. S. (2005). The gender similarity hypothesis. *American Psychologist, 60,* 581–592.

Jacobs, J. A., & Winslow, S. E. (2004). Overworked faculty: Job and stresses and family demands. *Annals of the American Academy of Political and Social Science, 596,* 104–129.

Levine, S. C., Huttenlocher, J., Taylor, J., & Langrock, A. (1999). Early sex differences in spatial skills. *Developmental Psychology, 35,* 940–949.

Levine, S. C., Vasilyeva, M., Lourenco, S. F., Newcombe, N. S., & Huttenlocher, J. (2005). Socioeconomic status modifies the sex difference in spatial skill. *Psychological Science, 16,* 841–845.

Lubinski, D., Benbow, C. P., Webb, R. M., & Bleske-Rechek, A. (2006). Tracking exceptional human capital over two decades. *Psychological Science, 19,* 194–199.

National Science Board. (2006). Salary differentials. In *Science and engineering indicators 2006* (ch. 3). Retrieved September 7, 2006, from http://www.nsf.gov/statistics/seind06/c3/c3s1.htm

Papierno, P. B., & Ceci, S. J. (2005). Promoting equity or inducing disparity: The costs and benefits of widening achievement gaps through universalized interventions. *Georgetown Public Policy Review, 10*(2), 1–15.

Pinker, S. (2002). *The blank slate: The modern denial of human nature.* New York: Viking.

Royer, J. M., & Garofoli, L. M. (2005). Cognitive contributions to sex differences in math performance. In A. M. Gallagher & J. C. Kaufman (Eds.), *Gender differences in mathematics* (pp. 99–120). New York: Cambridge University Press.

Shayer, M., Ginsberg, D., & Coe, R. (in press). 30 years on—a large anti-Flynn effect? The Piagetian test Volume and Heaviness norms 1975–2003. *British Journal of Educational Psychology.*

Spelke, E. S. (2005). Sex differences in intrinsic aptitude for mathematics and science? A critical review. *American Psychologist, 60,* 950–958.

Summers, L. H. (2005, January 14). *Remarks at NBER conference on diversifying the science and engineering workforce.* Retrieved July 25, 2006, from http://www.president.harvard.edu/speeches/2005/nber.html

Vasta, R., & Gaze, C. E. (1996). Can spatial training erase the gender differences on the water-level task? *Psychology of Women Quarterly, 20,* 549–567.

Voyer, D., Voyer, S., & Bryden, M. P. (1995). Magnitude of sex differences in spatial abilities: A meta-analysis and consideration of critical variables. *Psychological Bulletin, 117,* 250–270.

Wai, J., Lubinski, D., & Benbow, C. P. (2005). Creativity and occupational accomplishments among intellectually precocious youths: An age 13 to age 33 longitudinal study. *Journal of Educational Psychology, 97,* 484–492.

Wainer, H., & Steinberg, L. S. (1992). Sex differences in performance on the Mathematics section of the Scholastic Aptitude Test: A bidirectional validity study. *Harvard Educational Review, 62,* 323–336.

QUESTIONS FOR DISCUSSION
AND REFLECTION

1. Is the dearth of women in mathematically intensive fields due to cognitive factors (e.g., lower mathematical or spatial aptitude), to noncognitive factors (e.g., personal preferences and social–emotional considerations), or to some combination of the two? Why?

2. What is your opinion on the view that men are more serious about their careers than women are and use their training to pursue high-level jobs, while many women opt out in favor of family and relationships, working at lower level, part-time positions? Does this mean we should "face facts" and admit more men to doctoral programs, as some have suggested?

3. Why is it a problem to have gender imbalances in professions? Who is to say that a woman (or society as a whole) is better served if she enters a mathematically intensive field rather than, say, medicine, biology, or law?

4. Are tests such as the SAT—Mathematics biased against females because they underpredict their classroom grades in mathematics? Or, are males discriminated against because they have higher SAT—Mathematics scores than do females who earn the same classroom grades?

5. Do fewer females than males score in the outstanding region on mathematics aptitude tests because of brain differences?
6. If men are innately or genetically superior to women in academic aptitude, then why are there cultures in which there are no sex differences? Why are there cultures in which *women* have far higher aptitude test scores than American men?
7. Are sex differences in scientific ability and interest linked to hormone differences?
8. Did the male brain evolve to better process spatially complex information? Why or why not?
9. Are fewer women than men willing to devote 60 to 80 hours per week to becoming successful scientists because they place a higher priority on relationship and family issues? Why or why not?
10. Is it true that female graduate students are more likely to defer their career aspirations to those of their partners, moving for their partners even if it means taking less prestigious jobs, working part time, and so on? How many examples can you think of in which men moved for a woman's career?
11. Women do better than men in high school and college coursework, getting higher grades and graduating in higher numbers. Is this evidence that women have what it takes to be represented in science in greater numbers if they truly want to be in scientific and technical fields?
12. To what extent can we explain the lower numbers of women in science on the basis of negative stereotypes, including self-stereotypes, about women's ability and suitability?
13. Given women's greater drop-off after receiving higher degrees, are doctoral programs warranted in admitting more men because men are more likely to use their degree to its fullest?
14. Some have argued that women need institution-friendly policies, such as extended child-care leave. But might such policies negatively affect a field's progress if junior scholars worked part time for 5 years and then tried to reemerge as full-throttle scientists?
15. Who are the individuals who go on to become successful scientists? Do they represent the top 5% of aptitude, the top 1%, or, as some believe, the top one hundredth of 1%?
16. Should we cease trying to socially reengineer people's proclivities and accept that women and men have different aspirations? Why not let the two sexes pursue their different dreams?

AUTHOR INDEX

Numbers in italics refer to listings in the reference sections.

SUBJECT INDEX

Intelligence, and genetic factors, 118
Intelligence, general, 103, 105, 115, 123–124
Intelligence tests, 103, 105, 123, 232
Interests, 183
    and human development, 177–179
    and success in STEM careers, 87–88, 122, 150–151
Interest value, and subjective task value, 202–203
International Assessment of Educational Progress (IAEP), 133, 139
Intervention, "growing ability" (junior high), 51–52
Inzlicht, M., 51

Jacklin, Carol, 9, 14, 58
Jacobs, J. A., 219
Jacobson, L., 110
Japan, 29–30, 139, 224
Johns Hopkins University, 116
Junior high, 49, 51–52, 116

Kimura, Doreen, 8, 14, 215, 231, 233
Kucian, K., 194

Language ability, 165
Language style, 165
Laterality gradients, 191
Le Bon, Gustave, 102
Leeds University, 233n
Licht, Barbara, 48
Lifestyle values, 207
Linear regression technique, 218
Lippa, R., 87
Loenneker, T., 194
Long, J. S., 18
Longevity, of women, 196
Longitudinal studies, need for, 153, 193
Lubinski, D., 16, 42, 83–84, 88, 90, 216–217, 219, 229

Maccoby, Eleanor, 9, 14, 58
Magnetic resonance imaging (MRI), 115, 190
Male–male competition, 176–177, 184
Males
    with low early androgen levels, 150
    and math ability, 58–61, 116
    and mental rotation ability, 103, 125, 214
    and motor coordination, 193
    and object orientation, 57–58

    and spatial perception, 103–104
    in STEM careers, 93
    and visual–spatial abilities, 124–125
Male variability, 223
Mammals, hormonal effects in, 104–105
Manifest performance, 216
Map reading, 167
Martin, E., 194
Mason, M. A., 18–19
Massachusetts Institute of Technology (MIT), 17–18
Math ability, 41–42, 47–53, 58–61, 70, 104, 116
Math computations, 104
Math concepts, understanding of, 104
Math curriculum, 60, 134
Math performance, 133–134
Math reasoning, 59, 114
Math teaching, 52
Math tests, 133–134
Mechanics, 168
Medical schools, 121–122
Memory skills, 125
Menstrual cycle, female, 106
Mental rotation ability, 13, 42, 69, 74, 103, 106, 125, 167, 194, 214
Meta-analysis, 58, 103, 124, 126–127, 132–133
    on math performance, 133–134
    on spatial ability, 74–75, 134, 139
    on verbal ability, 134–135
Michigan Study of Adolescent Life Transitions, 206–209
Misuse of research findings, fear of, 122–123
Motivation, 64–65
Motor coordination, 193
Motoric systems, 167
Mozart effect, 74
Mozley, L. H., 195
Mozley, P. D., 195
Multiplier effect, 222
Murder, 164–165
Murray, C., 233n

National Academy of Sciences
    Committee on Science, Engineering, and Public Policy, 9
    "Maximizing the Potential of Women in Academe," 9
National Center for Education Statistics, *Trends in Educational Equity of Girls and Women*, 13–14

ABOUT THE EDITORS

Stephen J. Ceci, PhD, Helen L. Carr Professor of Developmental Psychology at Cornell University, has published extensively on topics related to group differences in achievement, intellectual development, and gap-closing. He is a member of the White House Commission on Children; the National Academy of Sciences Board on Cognitive, Sensory, and Behavioral Sciences since 1996; the National Science Foundation Advisory Board since 1997; and various National Research Council committees. He is a fellow of the American Psychological Association, the American Psychological Society, and the American Association for the Advancement of Science. Dr. Ceci founded and coedits the journal *Psychological Science in the Public Interest*, published by the Association for Psychological Science. In 2003 Dr. Ceci was the corecipient of the American Psychological Association's Lifetime Award for the Scientific Application of Psychology (with Elizabeth F. Loftus), and in 2005 he was the corecipient of the Association for Psychological Science's James McKeen Cattell Award for lifetime contribution (with E. Mavis Hetherington). He has published over 300 articles, chapters, books, and reviews, including several citation classics and 14 articles and books cited over 100 times each. In both 1997 and 2002 he was listed by Byrne and McNamara (*Developmental Review*) as one of the most cited developmental psychologists.

Wendy M. Williams, PhD, is professor in the Department of Human Development at Cornell University, where she studies the development, assessment, training, and societal implications of intelligence and related abilities. She holds a doctorate and master's degrees in psychology from Yale University; a master's degree in physical anthropology from Yale; and a bachelor's degree in English and biology from Columbia University, where she was awarded cum laude with special distinction. Dr. Williams cofounded and codirects the Cornell Institute for Research on Children, a National Science

ence Foundation–funded research- and outreach-based center that commissions studies on societally relevant topics and broadly disseminates its research products. She heads "Thinking Like a Scientist," a national education–outreach program designed to encourage traditionally underrepresented groups (girls, people of color, and people from low-income backgrounds) to pursue science education and careers. In addition to dozens of articles and chapters on her research, Dr. Williams has authored eight books and edited three volumes. Dr. Williams's writes regular invited editorials for *The Chronicle of Higher Education* and her research has been featured in *Nature*, *Newsweek*, *Business Week*, *Science*, *Scientific American*, *The New York Times*, *The Washington Post*, *USA Today*, *The Philadelphia Inquirer*, *The Chronicle of Higher Education*, and *Child Magazine*, among other media outlets. She served on the editorial review boards of *Psychological Bulletin*, *Psychological Science in the Public Interest*, *Applied Developmental Psychology*, and *Psychology, Public Policy, and Law*. Dr. Williams is a fellow of the American Psychological Society and of four divisions of the American Psychological Association (APA). In both 1995 and 1996 her research won first-place awards from the American Educational Research Association. She received the Early Career Contribution Award from APA Division 15 (Educational Psychology); the 1997, 1999, and 2002 Mensa Awards for Excellence in Research to a Senior Investigator; and the 2001 APA Robert L. Fantz Award for an Early Career Contribution to Psychology in recognition of her contributions to research in the decade following receipt of the PhD.